HYBRID PROJECT MANAGEMENT

HYBRID PROJECT MANAGEMENT

Cynthia Snyder Dionisio

WILEY

Published by John Wiley & Sons, Inc., Hoboken, New Jersey.
Published simultaneously in Canada.

For general information about our other products and services, please contact our Customer Care Department within the United States at (800) 762-2974, outside the United States at (317) 572-3993 or fax (317) 572-4002.

Wiley publishes in a variety of print and electronic formats and by print-on-demand. Some material included with standard print versions of this book may not be included in e-books or in print-on-demand. If this book refers to media such as a CD or DVD that is not included in the version you purchased, you may download this material at http://booksupport.wiley.com. For more information about Wiley products, visit www.wiley .com.

Library of Congress Cataloging-in-Publication Data Applied for

Paperback ISBN: 9781119849728

Cover Design: Wiley
Cover Images: © Botond1977/Shutterstock;
BNMK 0819/Shutterstock; iam2mai/Shutterstock

Set in 11/13pt and Helvetica Neue LT Std by Straive, Chennai, India

SKY10039868_121322

Contents

Acknowledgments

How do you express gratitude for all the people over 25+ years who have contributed to your knowledge and skills in project management? In my years of practicing, teaching, and writing about project management I have come across thousands of practitioners and professionals. While there are far too many to note, I would be remiss if I did not acknowledge a few people who have helped me in my career and with this book.

I have learned more about project management by teaching than I did by practicing. It is in the teaching that I have improved my own skills. Thus, I am very grateful for the thousands of students who allowed me to contribute to their knowledge and careers, as it has helped me become a better project leader and course facilitator.

Of course, I could not reach students without the organizations that hired me to teach. In particular I would like to acknowledge my friend and colleague Teidi Tucker at Management Concepts for a personal and professional relationship that has spanned over 20 years. You are such a delight!

I am grateful to Amy Sell, Dianne Starke, and all of the wonderful people at LinkedIn. I learned a lot about hybrid project management by developing a course for LinkedIn Learning. Thank you for that opportunity!

LaVerne Johnson and Amy Gershen with the International Institute for Learning were some of the first people I worked with in a training environment and are wonderfully supportive in my current role as Practice Lead. I appreciate you both.

Over my career I have spent thousands of hours volunteering with the Project Management Institute (PMI). I am so very grateful for the leadership roles PMI has entrusted to me. Past and present friends and colleagues include Stephen Townsend, John Zlockie, Dani Ritter, Kristin Vitello, Marv Nelson, Barbara Walsh, and Roberta Storer. The teams that guided the fourth, sixth, and seventh editions of the *PMBOK® Guide* are some of the best professionals in the business. Each person expanded my knowledge and understanding of the full range of project management. I have to give a special shout out to Larkland Brown. His willingness to read parts of this book and provide sage advice have definitely improved the quality of the content.

The professionals at Wiley are among the best in the business. I am so grateful to have published numerous books with Wiley and the For Dummies folks over the years. Kalli Schultea and Amy Odum have been a delight to work with on this and many other books.

It would not be possible to thrive in this profession without the support of my loving family. Pad, Mombat, and Bunny, you are all the best and always in my heart.

Introduction

Project managers make progress, change, new ideas, new technologies, and breakthroughs possible. We are a part of every profession and have proven our value through our leadership, ingenuity, courage, and discipline. We have been delivering value to organizations, governments, military, and nonprofits since before there was a profession called project management.

In the 1980s and 1990s project management was primarily linear and process driven. There was a heavy emphasis on planning, managing, and controlling project work. We were process driven and documentation heavy. Staying true to the plan, limiting scope changes, and adhering to baselines were the primary means by which we executed and controlled our projects. This approach has come to be known as a waterfall approach because we managed our projects one phase at a time in a linear fashion.

By the late 1990s it was becoming clear that this approach worked in some situations, such as building a bridge, but was a recipe for failure in other situations, such as developing software. In 2001 a group of 17 software developers met at a mountain resort and came up with a new approach that was based on four values and 12 principles. They recorded the values and principles in a document called the Agile Manifesto.

The Agile Manifesto moved away from heavy up-front planning and embraced evolving scope, collaborative working relationships, and servant leadership. The practices, mindset, and overall approach to creating deliverables was 180 degrees from the waterfall methods.

As with most new approaches, some people embraced the practices and mindsets wholeheartedly. In fact, some practitioners became rather evangelical about the Agile mindset and were termed Agilistas. Other practitioners were more along the lines of "Agile in name only." In other words, they used the Agile terminology but did not fully embrace the mindset. Today the majority of the Agile practitioners conform to Agile practices but perhaps aren't as fanatical about their implementation as the Agilistas.

Twenty years after the Agile Manifesto was first developed, there are many practitioners who find value in the waterfall approach and the Agile approach. These practitioners recognize there are many variables that determine which method to use and that there is rarely a need to be on one end of the spectrum or another. They see significant value in embracing both approaches, using waterfall techniques for some deliverables and Agile techniques for other deliverables. We call this "hybrid project management."

Hybrid project management is about being flexible enough to assess the project, deliverables, environment, and stakeholders to determine the best means to achieve the intended outcomes. Many hybrid projects use a waterfall framework at a high level and apply Agile approaches to specific deliverables as appropriate. It is my opinion that while the servant leadership practice that is a hallmark of Agile is not fully utilized in all hybrid projects, the practice of engagement, collaboration, and facilitation is more prevalent than the command-and-control perspective that was common in the 1980s and 1990s. Thus, we see the industry moving away from the far ends of the project management spectrum and more toward a practice that is inclusive of the best practices from each approach.

This book is intended to present a variety of ways to deliver projects. Whether you are a new practitioner or someone with decades of experience, it is my hope that you will find some new ways of practicing our discipline and discover some new techniques you can apply on your projects.

1

Introducing Project Management

As professional project managers it is no longer enough to deliver results that conform to requirements and are on time and on budget. Our role in driving change and transformation, developing new products and updating existing products, creating new technologies and finding ever-better ways of doing things has evolved. Now we need to be more business savvy, market responsive, and aware of how our profession is changing and progressing.

One of the most significant changes in our profession is the recognition that as professionals we must understand and embrace different ways of delivering value. After all, the whole purpose of projects is to bring value to stakeholders, whether that value is via a new product, a new service, or more efficient processes. Different project deliverables require different approaches and techniques. To excel in our role, we need to know our options for delivering value and understand the variables that determine the best fit for each deliverable.

Value: Something of worth or importance.

Deliverable: A component or subcomponent of a product or service. A deliverable can be stand alone, or part of a larger deliverable.

In this chapter we will describe four ways of creating deliverables and define terms associated with each approach. Next, we will identify the variables you need to consider in order to select the best development approach for each deliverable in your project.

THE SPECTRUM OF DEVELOPMENT APPROACHES

Development approach: The means by which the project team will create and evolve deliverables.

A development approach refers to how the project team creates and evolves deliverables. Some development approaches emphasize understanding all the requirements before designing a solution and then creating the deliverables based on the solution design. Other development approaches start with a bare-bones deliverable and evolve the solution based on feedback. They are different approaches to creating project deliverables.

Warning! A development approach is not a life cycle. We will cover life cycles in Chapter 5.

In this chapter we will describe four development approaches:

- Waterfall;
- Incremental;
- Iterative; and
- Agile.

A waterfall development approach is what we call predictive. In other words, we like to be able to predict the schedule and budget based on stable scope. Incremental, iterative, and Agile development approaches are adaptive, which means they are flexible enough to allow changes in requirements and scope.

Adaptive: An approach for creating deliverables that allows for uncertain or changing requirements.

Predictive: An approach for creating deliverables that seeks to define the scope, schedule, and budget toward the beginning of the project and minimize change throughout the project.

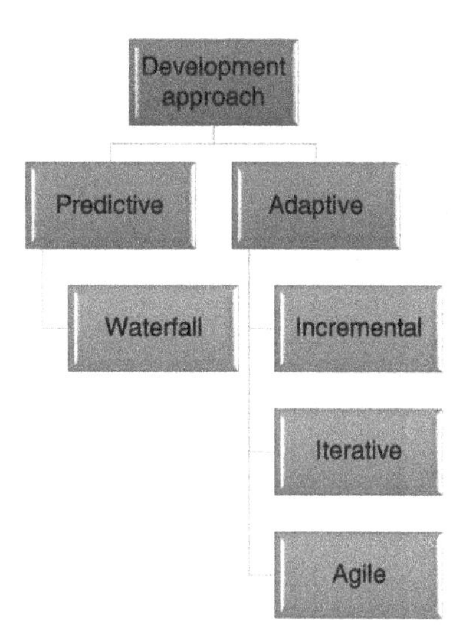

FIGURE 1-1 Development approach.

Waterfall

Waterfall: A predictive approach for creating deliverables that follows a linear pattern of completing one phase of work before starting the next one.

A waterfall development approach is predictive in nature. In other words, it starts with well-defined scope, which the project team then progressively elaborates into greater levels of detail. The work is then sequenced, duration and cost estimates are developed, and eventually a baseline is set. Throughout the project, progress will be measured against the baselines. With a waterfall approach the project manager endeavors to keep change to a minimum and follow the project plan.

Figure 1-2 shows a life cycle for a light rail project that would use a waterfall approach. You can see how one phase completes before the next one begins, and the shape of the graphic looks like a waterfall. In the Environmental Impact Analysis phase, studies on expected site impact, materials analysis, geological surveys, life cycle assessment, and similar work would be conducted. In the Plan phase, detailed resource, budget, schedule, communication, risk, and other plans would be developed. At the end of the Plan phase those plans would be baselined. The Engineering phase would be comprised of blueprints, architecture, modeling, and other similar work to ensure the designs meet the needs, are compliant with regulatory requirements, and are minimally disruptive to the environment. The Construction phase is

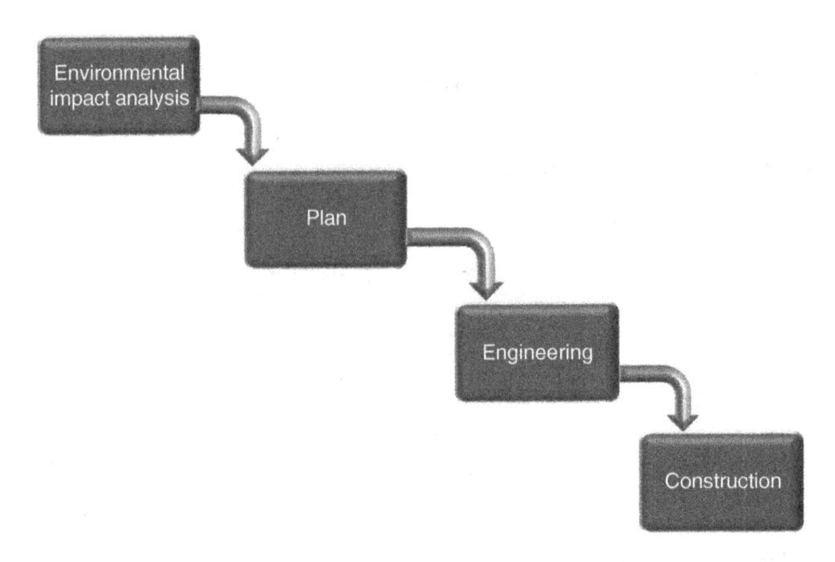

FIGURE 1-2 Waterfall approach.

where all the physical work is carried out. It is the most visible, uses the most budget, and likely takes the longest. Progress in the Engineering and Construction phases would be compared to the baselined plans to ensure the project stays on schedule and on budget.

A waterfall approach is best when the requirements can be defined up front and when the scope of the project is not expected to change. This approach is often used for projects with large budgets, where detailed planning can help reduce uncertainty and risks. Projects that have high-risk deliverables or have significant regulatory oversight are also a good fit for a waterfall approach.

Types of projects that use a waterfall approach include:

- Construction;
- Defense projects such as building a new aircraft, ship, or tank;
- Medical devices; and
- Infrastructure including roads, bridges, or mass transit.

Iterative

An iterative approach is adaptive in nature. It is used when there is a high-level understanding of the desired outcome, but the best way to achieve that outcome is not defined. The project team uses a series of iterations to get clarity on the best method to deliver results.

Iterative: An adaptive development approach that begins with delivering something simple and then adapts based on input and feedback.

Iteration: A brief, set time interval in a project where the team performs work. Also known as a timebox or sprint.

An iterative approach could be used for designing a new multipurpose bicycle. It might start with an idea on a drawing board that can be shown to key stakeholders for feedback. Once the stakeholders are happy with the design, the team may use cheap materials to build a cheap frame mock-up that people can look at and sit on to provide more feedback. Once the frame shape is settled, the next iterations can focus on finding the right materials. The right materials affect the ride, price, weight, handling, and expected life span.

When the frame size and materials are decided on, the team can conduct iterations to determine the best gears, brakes, and other componentry. Only when the team has incorporated all the relevant feedback will they finalize the design, materials, and specifications so they can go into production.

Figure 1-3 shows a generic example of an iterative approach. Notice that each iteration provides information to the next iteration. The number of iterations depends on the feedback and when the decision makers agree that the final iteration will meet the objectives of the project.

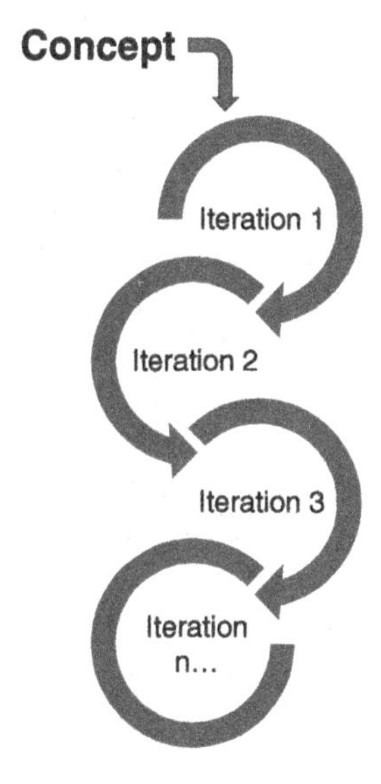

FIGURE 1-3 Iterative approach.

An iterative approach may be used in conjunction with Agile methodologies, especially for software development; however, there are many other usages in addition to software development. Types of projects that could use an iterative development approach include:

- New product development;
- Software development; and
- Marketing campaigns.

Incremental

Incremental: An adaptive development approach that begins with a simple deliverable and then progressively adds features and functions.

An incremental approach is adaptive in nature. It is used when the end product can be decomposed into smaller components and deliverables can be deployed incrementally. Each increment learns from previous deployments and adds or improves features and functionality of the deliverables.

An incremental approach might start with an idea and build a basic version of the idea and release it. After release, the team would gather feedback, such as how people use the product, which features they use the most, which features they don't use, and the number of calls for support. This feedback informs the next increment for release. Depending on the product, the team may add a new component or might upgrade software.

Minimum viable product: The first release of a product that contains the least number of features or functions in order to be useful.

This approach could be used to develop an online learning course. The first increment could include slides that can be accessed online and a PDF document that can be downloaded. These two elements might be what is called a minimum viable product. In other words, it has just enough features that people will buy it. Some customers may provide explicit feedback on the product. However, because the product is online, customer behavior can be monitored to indicate how much time they spent with each feature, which ones they returned to, and when they logged out.

Based on feedback, the next increment could include built-in exercises, quizzes, and interactive activities. This would be released, and more data would be collected. The next increment might include videos, audio clips, or threaded discussions. Development and upgrades would continue until a decision was made that the product was complete.

Notice with this incremental approach that the team releases a product that is complete with each increment. They don't have to wait until the whole product is done or integrated before it is released. This allows the team to learn rapidly and update their plans based on stakeholder feedback.

Figure 1-4 shows a generic example of an incremental approach. This example shows four increments, where each increment would add more functionality.

An incremental approach is often used with Agile methodologies for software development, though that is not the only use for an incremental approach. Types of projects that could use an incremental development approach include:

- Customer loyalty programs;
- Application development; and
- Online learning.

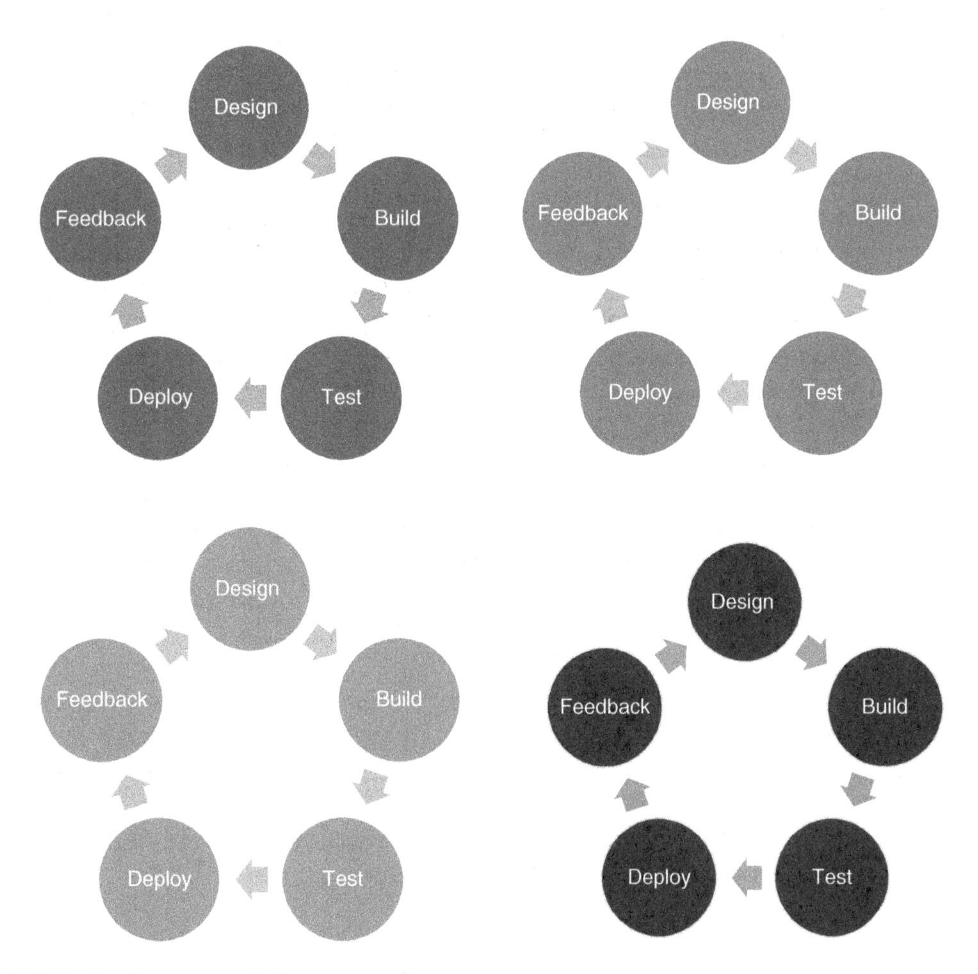

FIGURE 1-4 Incremental approach.

Agile

Agile: An adaptive way of delivering value by following the four values and 12 principles established in the Agile Manifesto.

As mentioned in the introduction, Agile is a mindset based on values and principles. There are several frameworks or methodologies that incorporate those values and principles. They all include iterative development and continuous feedback. This book won't espouse one methodology over another but rather will address Agile as an approach for developing deliverables.

Iterative and incremental approaches are used with Agile; however, the iterations or timeboxes are very short, usually one, two, or four weeks in duration. At the end of every iteration (sometimes called a timebox or a sprint), the team demonstrates the work they have accomplished to key stakeholders. Stakeholders provide feedback, and a backlog of features and functions is then prioritized for the next iteration.

Agile approaches have several unique aspects to them that will be described throughout this book, such as different roles, meetings, prioritization methods, and scheduling.

An example of using an Agile approach could be a county that wants to understand how its residents are using its parks and open spaces. They could build an application that pulls data from online searches, educational programs, surveys, parking meters, vendors, and other data sources. The application would compile data from these various sources and make it searchable, create tables, charts, dashboards, and other tools. This information could help the county for staffing, planning, and resident satisfaction.

The team would start with this high-level concept and a list of features and functions the customer has asked for. The customer would prioritize the work, and the team would determine which of the prioritized features they could get done in the coming iteration. At the end of the iteration they would demonstrate their work, receive feedback, and move onto the next iteration. At some point, they would have enough features and functions built that they could release the application for use. If needed, they could add more functionality in later releases.

This example could use iterative practices to evolve the various aspects of the application, such as the dashboard behavior. It could also use incremental practices to release some of the functionality, do more work, release more functionality, and so forth.

Figure 1-5 shows a generic example of an Agile approach. Each sprint uses feedback from the previous sprint to plan and develop the upcoming sprint.

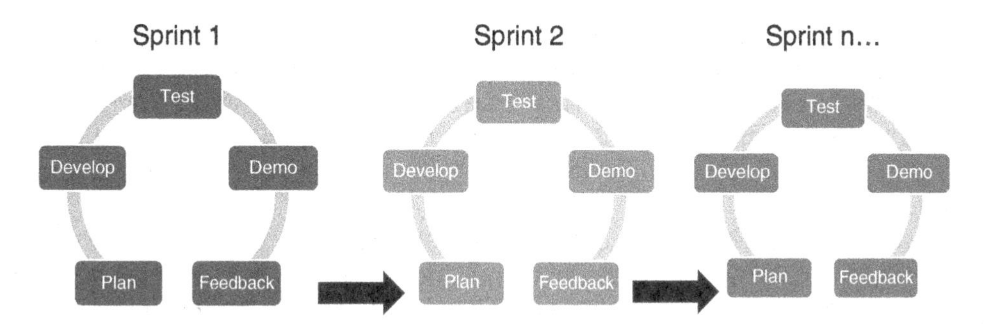

FIGURE 1-5 Agile approach.

HYBRID PROJECT MANAGEMENT AND DEVELOPMENT APPROACHES

A hybrid approach uses some predictive and some adaptive approaches. Expanding on the iterative example of developing a new bike, the design of the bike could use an iterative approach and then when preparing for manufacturing and later distribution, they could use a waterfall approach. The design aspect of the bike uses feedback to ensure the bike is meeting the needs of potential customers. The manufacturing and distribution require up-front planning and a stable set of requirements.

Hybrid project management: A blend of predictive and adaptive approaches to delivering value, determined by product, project, and organizational variables.

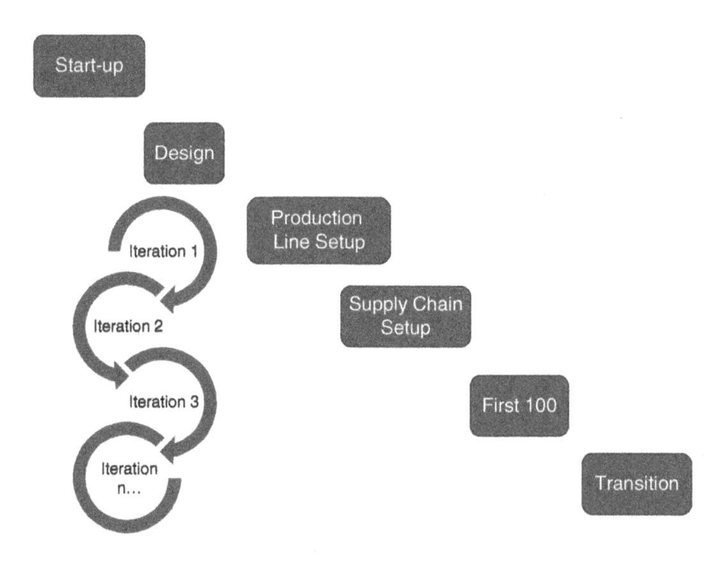

FIGURE 1-6 Hybrid approach 1.

Developing a new sports watch could use an iterative approach for the software part of the watch and waterfall for the hardware part of the watch. As people use the watch, they can make requests for new or modified features, which can be deployed as updates to the watch operating system, but the physical watch itself won't change. Figure 1-7 shows a waterfall approach that includes outsourcing the manufacture of the watch.

Finding the right contractor and going through the contracting legal work is part of the project. Once it moves to manufacturing, the hardware part of the project will be complete. As the work for the watch hardware is happening, the team can incrementally develop features and functions for the watch. By the time the manufacturing is ready to begin, the software should have gone through a few iterations and be ready for deployment.

The different development approaches usually work well with specific practices and ways of working. This book will point those out, but the beauty of hybrid project management is that you can tailor, mix, and match to meet the needs of your project, environment and stakeholders.

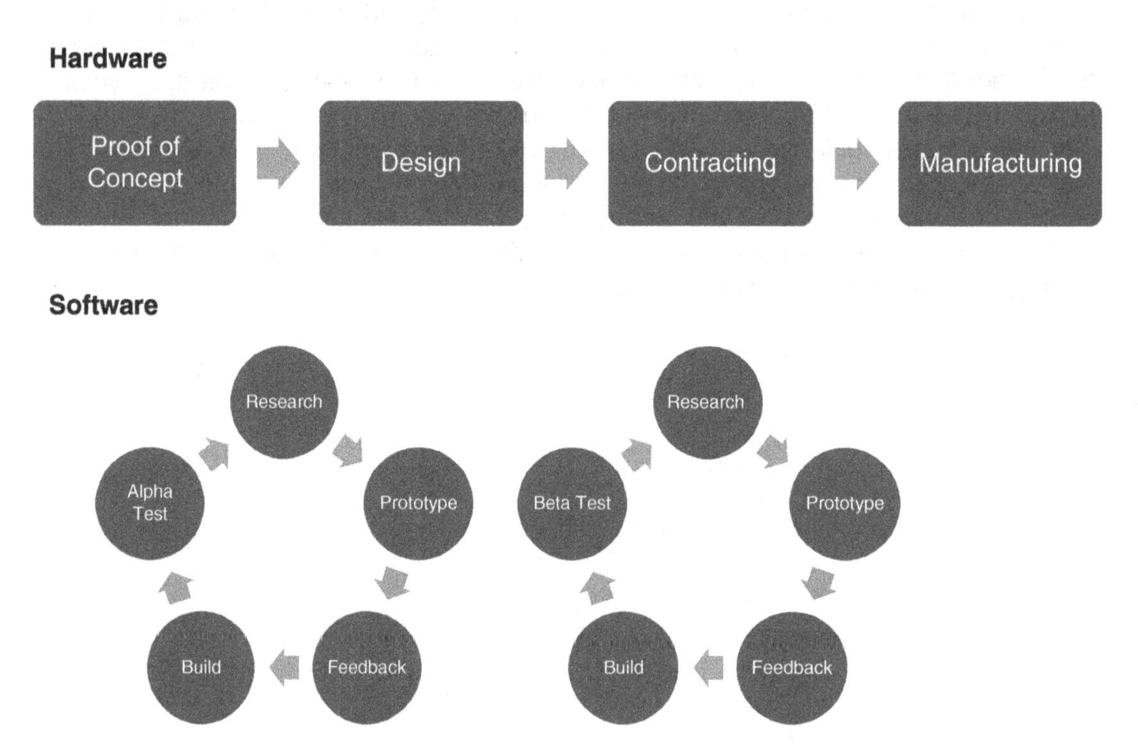

FIGURE 1-7 Hybrid approach 2.

SUMMARY

In this chapter we introduced key concepts and terminology for hybrid project management. We looked at four different ways to create deliverables:

- Waterfall;
- Iterative;
- Incremental; and
- Agile.

We also looked at hybrid project management as a way to combine or mix and match those development approaches to meet the needs of the project.

Key Terms

adaptive
Agile
deliverable
development approach
hybrid project management
incremental
iteration
iterative
minimum viable product
predictive
value
waterfall

2

Choosing a Development Approach

Choosing a development approach for deliverables in your project requires familiarity with the various options (waterfall, iterative, incremental, and Agile), an understanding of the product, and contextual information about the project and organization. While there isn't a neat and precise way to come up with the perfect approach for each deliverable, there are some guidelines that can help you evaluate the right approach for your project.

In this chapter we will look at how product variables, project variables, and the performing organization can influence the selection of a development approach.

PRODUCT VARIABLES

It makes sense to start with the product variables since these relate to the scope and outcomes the project will deliver. We'll review eight product variables to consider when evaluating the best development approach for each deliverable:

- Innovation;
- Scope stability;
- Requirements certainty;
- Ease of change;

- Risk;
- Criticality;
- Safety; and
- Regulatory.

With each variable, I'll describe ways a hybrid approach can be used.

Innovation

Innovation takes into consideration the degree to which the technology and methods you will use on the project are new and untested versus known and standardized. Using methods and processes you are familiar with is conducive to waterfall approaches. Cutting-edge technology or experimental processes work better with adaptive approaches.

A project to repave eight neighborhoods does not require any innovation. The technology and methods are well known, so a project like this works well with a waterfall approach. Conversely, a project to build a battery that can last for 10 years in 0 gravity requires significant innovation. Therefore, this type of project would work well with iterative and incremental approaches. The team would require a lot of creativity and the ability to experiment and try different ways to achieve the intended result.

Hybrid options: A hybrid approach is good if you have some deliverables that are known and some that are newer. You can also use adaptive methods until you have tested the technology and are comfortable with it and then move to processes that support a known technology.

Scope Stability

How likely is your customer to change their mind, add new features, or request something different? If you are working on a project where the scope is fixed and unlikely to change, such as installing landscaping in a housing development, you can use a waterfall approach. In contrast, if your customer is fickle or has a lot of new ideas they want to try out, such as rebranding a product line, then you should consider one of the adaptive approaches.

Hybrid options: You may be working on a project where some deliverables are stable and some are subject to change. In these circumstances the flexibility of a hybrid approach is a good choice. Another option is to use adaptive methods until the scope has stabilized and then implement more of a waterfall approach.

Requirements Certainty

Requirements certainty is related to scope stability, but it is a bit different. The scope is what you are delivering, the requirements are the capabilities that must be present and conditions that must be met to achieve the project objectives.

Requirement: A capability that must be present or a condition that must be met to achieve the project objectives.

Some projects have very clear requirements from the start, for example, install a three-story parking garage that can hold 500 cars. Clear requirements lend themselves to waterfall approaches.

Many projects don't know all their requirements at the start. The team expects the requirements to evolve and new requirements to be added throughout the project. A project to establish a concierge service for a high-end credit card might start out with some high-level concepts and ideas, but as the service is rolled out, those requirements might evolve and change based on user requests and feedback.

Hybrid options: Using an adaptive approach to test different requirements or requirement sets is a good way to start a project when the requirements are uncertain or subject to change. Once there is more certainty, you can transition to more of a waterfall approach. You can also document and manage requirements that are certain, while using adaptable methods to stay flexible with those that could evolve.

Ease of Change

Change is a way of life, especially on projects. But not all projects absorb change easily. A project to create an electronic performance dashboard can absorb changes in scope or requirements fairly easily. This type of project fits well with an adaptive development approach.

A project to build a bridge does not respond well to change. For this type of project, you want to make sure you have all the specs correct before you start construction because any change could be very time-consuming and costly! Therefore, you would want to use a waterfall approach where you lock in your scope and designs prior to starting construction.

Hybrid options: To address projects where some deliverables are easy to change and some aren't, you can split out those deliverables that are easy to change and manage them using adaptive approaches and manage those that are not easy to change with a rigorous change control approach that is the hallmark of waterfall approaches. Another option is to make decisions and allow changes as late in the project as possible and then lock down the product so no more changes can occur after a certain point in time.

Risk

The type of risks on a project indicates the most appropriate risk responses. Risk responses for risks associated with product acceptance or new technology can include using an adaptive development approach. An adaptive approach allows the team to experiment and develop prototypes and then evolve the product based on the outcomes and feedback.

Risk: An uncertain event or condition that can have an impact on a project.

For projects with risks associated with safety or where you can't fix something once it is complete, a waterfall approach is best. For example, if you are launching a satellite, once it is launched you can't redo or rework something; therefore, the up-front planning and robust risk management that is used in a waterfall approach is best.

Hybrid options: Risk management and responses are necessary regardless of the approach used; however, the type of response can vary. Therefore, a hybrid project will have a variety of options for responding to risk. The rigorousness of the risk management process can flex depending on the type of risks that are present in the project.

Criticality

Criticality deals with the relative importance of a component, deliverable, or project. For example, there is a high degree of criticality with maintaining the power for hospitals. A component or deliverable with a high degree of criticality usually indicates a waterfall approach is best. A component that is easy to replace and does not have a significant effect if it fails may work with an adaptive method. For example, a project to implement an online ordering system would have a requirement to send an automated email confirming the order. If this function fails, it is relatively easy to fix, and no one will be harmed if they don't get the confirmation email.

Criticality: The importance of a component, deliverable, or project.

Hybrid options: If you have a variety of deliverables or components and some are critical and some aren't, determine at the outset which deliverables require the degree of planning, testing, and documentation necessary for critical deliverables. For those deliverables that are not critical, you can use less intensive processes.

Safety

When safety issues are involved, most projects rely on a waterfall approach. For example, projects that develop implantable medical devices have significant safety concerns. They require the planning, documentation, and testing that is common in waterfall projects. Conversely, a project to update a gaming application on a smartphone doesn't have a lot of safety issues, and so an adaptive approach would work well.

> **Hybrid options:** Usually not every deliverable for a project has safety impacts. For those deliverables that clearly have no safety impacts, you can use an adaptable approach while maintaining robust planning, documentation, and testing for those deliverables with safety concerns.

Regulatory

Many projects are done to achieve or maintain regulatory compliance. This can include plant inspections for facilities that work with hazardous materials or educational institutions that need to maintain compliance with accreditation requirements. Most regulatory agencies want to see detailed documentation and rigorous policies and procedures that demonstrate compliance. These projects use waterfall approaches. Where there isn't a need to demonstrate compliance or alignment with regulatory agencies, both waterfall and adaptive approaches are valid.

> **Hybrid options:** In the event that only some aspects of a project have regulatory considerations, separate those aspects with regulatory concerns from those without. Invest in the required amount of policy, process, and documentation for those deliverables with regulatory requirements and relax the processes and documentation for deliverables without regulatory requirements.

PROJECT VARIABLES

Project variables that influence the development methods include:

- Stakeholders;
- Delivery options; and
- Funding availability.

Stakeholders

Projects have a wide range of stakeholders, some of which will be well known to the project team, such as the sponsor or product owner, and some that the team will never meet, such as certain end users or members of the public. One of the strengths of Agile methodologies is the access to key stakeholders. In a purely Agile environment, key stakeholders, such as the product owner, are available to the team to answer questions and protect the team from interference. They review work at regularly scheduled intervals, such as demonstrations every two weeks, and they prioritize features in a backlog. Thus, for projects that use Agile methods to develop products, access to key stakeholders is a necessity.

Projects that use waterfall methods usually have less need of and less access to stakeholders. Obviously, there is some interaction, but it is not as consistent or frequent as in Agile projects.

Hybrid options: If you are working on a project where some deliverables are using adaptive methods, you will want to stay in close contact with the appropriate stakeholders on a regular basis. For those aspects of the project that are developed with waterfall approaches, a monthly status report should be sufficient. You can apply hybrid methods by summarizing the adaptive information in the monthly status report that goes to management.

Delivery Options

Does your project have one main deliverable, or can it be decomposed into multiple smaller deliverables? Do all the deliverables have to be released at the same time, or can they be released in batches? The answers to these questions will point you in the right direction for choosing a development approach.

Typically projects with one delivery at the end use a waterfall approach so you can plan the development, testing, and delivery. A project to build a new hotel is an example of a project with one main release. Projects where you can work on several different deliverables but bring them together before release work well with an iterative approach. A project to update a payroll system might have multiple components, but they all have to be integrated before release. Projects that have periodic deliveries, such as a website, work well with incremental approaches.

Hybrid options: For projects with multiple deliveries that are developed using adaptive methods along with some deliverables that use waterfall methods you can support the adaptive development by having a scrum master work with the adaptive team. The scrum master can liaise with the project manager for the overall project and provide status information, milestones, and delivery dates, which can be incorporated into a waterfall framework and schedule.

Funding Availability

Many projects that are involved with new product development, especially digital products, will start with a small budget, and as the product gains market share and becomes profitable, they will add features and functions. This business model essentially uses the profit from the product to fund future upgrades and enhancements. This model is often used with Agile projects or projects that use iterative approaches. The up-front investment is comparatively minimal, and if the product doesn't do well, the project is cancelled with minimal loss of investment. This model can also be used if there is uncertainty about the amount or timing of funds. An Agile approach allows the team to deliver some value for the investment, even if not all the objectives are realized.

Multiyear projects with large budgets need to consider funding availability. Some projects are constrained by fiscal year funding, necessitating planning around the availability of funds. If there is only one large deliverable at the end, a waterfall approach is often used, and work is planned around funding availability.

> **Hybrid options:** The most common hybrid scenario for this variable is when a new product idea is envisioned and the market is tested using adaptive methods. If/when the decision is made to move forward and more stable funding is available, the development and delivery can be scaled up using waterfall methods.

ORGANIZATION VARIABLES

Organizational variables should match with the preferred development method, or the project will struggle to be successful. Four organizational variables to pay attention to are:

- Structure;
- Culture;
- Project team; and
- Experience and commitment.

Structure

There is a broad spectrum of organizational structures that range from hierarchical to matrix to flat. A hierarchical structure has lots of layers of management, and often the functions are siloed. This type of structure usually has a robust set of policies and procedures, and the communication and interaction between functions may be limited. Projects that use a waterfall approach do better in structures like this than Agile projects. The exception is if a particular function, or group of functions, such as IT, employs Agile methods.

Organizations that are flatter and more responsive to change and adaptation are where adaptive project management practices thrive. There are less layers of bureaucracy that require approval, and development is usually more nimble.

Hybrid options: Agile methods don't work well in bureaucratic environments. However, where it makes sense, you can still use some of the methods and techniques found in Agile projects, such as daily stand-ups and task boards.

There are also some organizations that employ Agile practices in some disciplines (such as IT), and other disciplines are primarily waterfall. If your project has team members that are used to different development approaches, you can blend approaches and educate your team about the benefits of each development approach.

Culture

Company culture is a big determinant in choosing the best way to manage a project. Projects that use Agile methods are based on a climate of trust, where the team is empowered to make many of the decisions about the project. They establish their ways of working, communicating, troubleshooting, and so forth. Agile teams are often self-managing, without a specific designated leader. This way of working will not work in a highly bureaucratic or "command and control" environment.

Conversely, a project that requires a lot of documentation, sign off on decisions, rigorous processes, and so forth, will do well in a hierarchical environment but will not fit as well in a flatter organization. One of the reasons many organizations struggle or fail in implementing an Agile methodology is that the company culture does not match with Agile ways of working.

Hybrid options: Unless you are working on an organizational transformation initiative, changing the company culture to adopt different ways of working on projects is likely not part of your project scope. However, you can introduce a few techniques from different approaches. For example, in a waterfall environment you can schedule tasks for prototyping and experimentation. In an Agile environment you can include milestones, status reports, and maybe even talk about a critical path for a release. By infusing a few practices from different development methods you can soften resistance to different ways of working and help people see that different approaches can be productive.

Project Team

The size and location of project team members influence the development approach. Agile methods work best with teams that are 5–10 people who are collocated (in the same space). While you can apply Agile methods with larger groups, or with virtual teams, it is more challenging. Agile teams have daily stand-up meetings, and much of their collaboration takes place in one-on-one conversations—preferably with a whiteboard nearby to capture ideas. These practices are more challenging when you have more than 10 team members or if team members are in different locations, especially if they are in different time zones or different countries.

Large teams work better with the structure of a waterfall project. There are fewer meetings, much of the documentation is electronic, and dispersed team members are not uncommon.

Hybrid options: For a large team that wants to use Agile practices, you can adopt one of the Agile at Scale methodologies, such as Scaled Agile Frameworks (SAFe), Large Scale Scrum (LeSS), or Scrum at Scale (SaS). You can also back off some of the practices that are more challenging with a large or dispersed team, such as daily stand-ups. Another option is to move much of the data that is usually visible in the team room, such as task boards and burn charts to an electronic location.

On large projects with multiple deliverables, you may find it useful to manage the overall project using a waterfall framework, while some team members use adaptive processes and others use waterfall practices.

Experience and Commitment

The final variable is experience and commitment—both on the part of the team and the organization. Teams and organizations with experience and commitment to a particular method of managing projects can struggle with adopting new ways of working. Even if the team has experience and commitment to working one way, if the organization isn't aligned, it will be a rough ride. On the other hand, if an organization wants to try a new way of working, such as investing in Agile practices, but the team is not on board, it will be a struggle.

Hybrid options: As with the other variables in the organization category, a good option is to introduce alternative practices a bit at a time to let the team and the organization get used to them and see that they are effective. You can also point out the similarities between practices; for example, a retrospective used in Agile is similar but not the same as a lessons learned used in waterfall approaches. You can blend the practices for lessons learned and retrospectives to blur the lines. You can also blend some of the estimating practices, such as using the estimates from planning poker in a multipoint estimate or use planning poker to come up with an estimate in a waterfall project. By taking small steps and proving their effectiveness, you can open the door to a more hybrid environment.

DEVELOPMENT APPROACH EVALUATION TOOL

Now that we have reviewed the variables that influence the development approach, we need to find a simple way to evaluate a project or project deliverables against the variables to determine the best approach. A good place to start is by rating your project or deliverable for each variable on a scale of 1 to 5.

The rating scales below are set up so that a rating of "1" is a better fit for a waterfall approach and a rating of "5" fits better with an Agile or adaptive approach. These ratings are just an example of what you might want to use; your project or organization may have different needs, so tailor these scales to fit with your environment.

Product Variables

Innovation

1	3	5
Stable technology	Newer technology	Technology doesn't exist yet

Scope stability

1	3	5
Stable scope	Some changes expected	Evolutionary

Requirements certainty

1	3	5
Clear requirements	Some requirements known	Unknown requirements

Ease of change

1	3	5
Change puts project at risk	Change requires rework	Easy to change

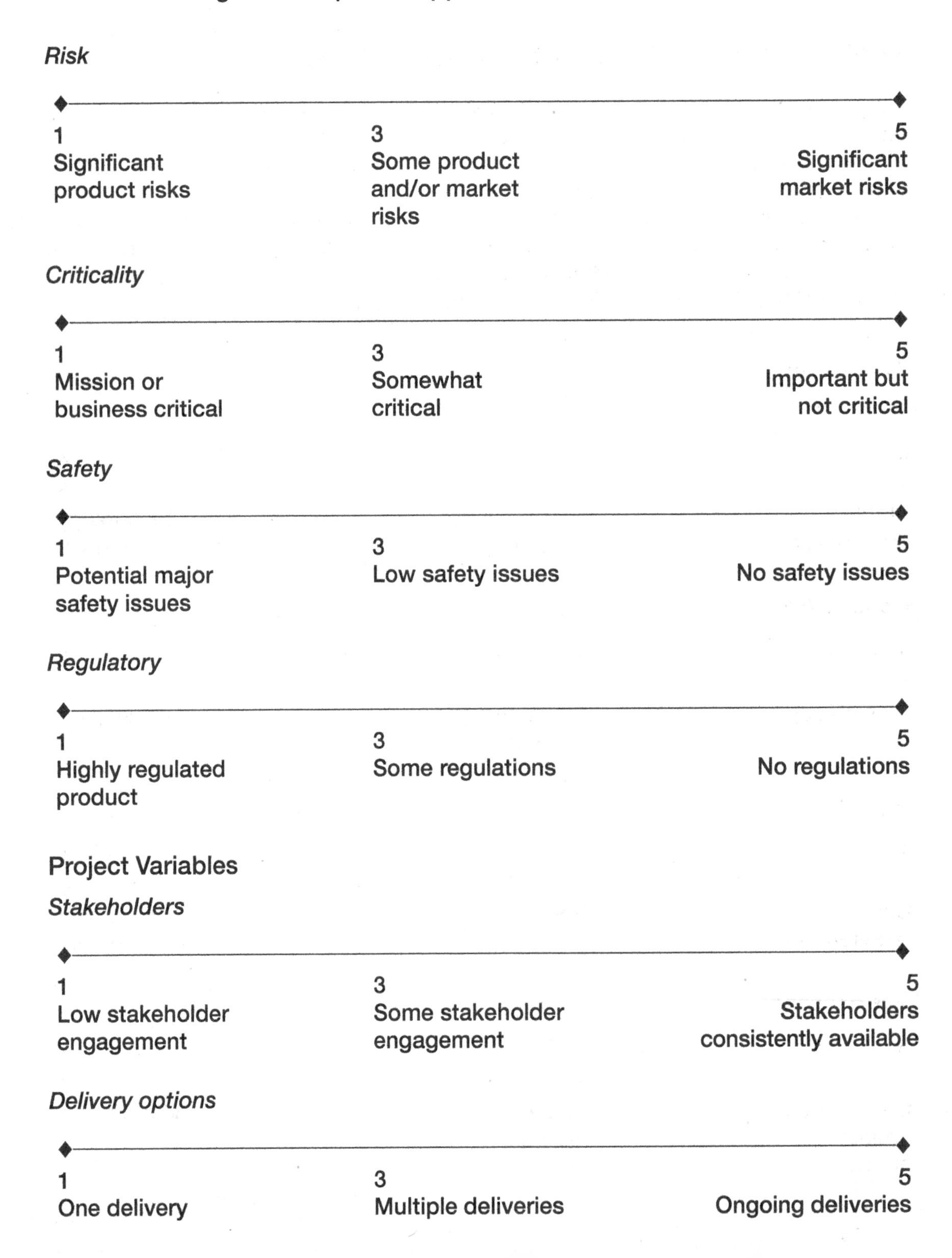
Risk
1
Significant product risks
3
Some product and/or market risks
5
Significant market risks
Criticality
1
Mission or business critical
3
Somewhat critical
5
Important but not critical
Safety
1
Potential major safety issues
3
Low safety issues
5
No safety issues
Regulatory
1
Highly regulated product
3
Some regulations
5
No regulations
Project Variables
Stakeholders
1
Low stakeholder engagement
3
Some stakeholder engagement
5
Stakeholders consistently available
Delivery options
1
One delivery
3
Multiple deliveries
5
Ongoing deliveries

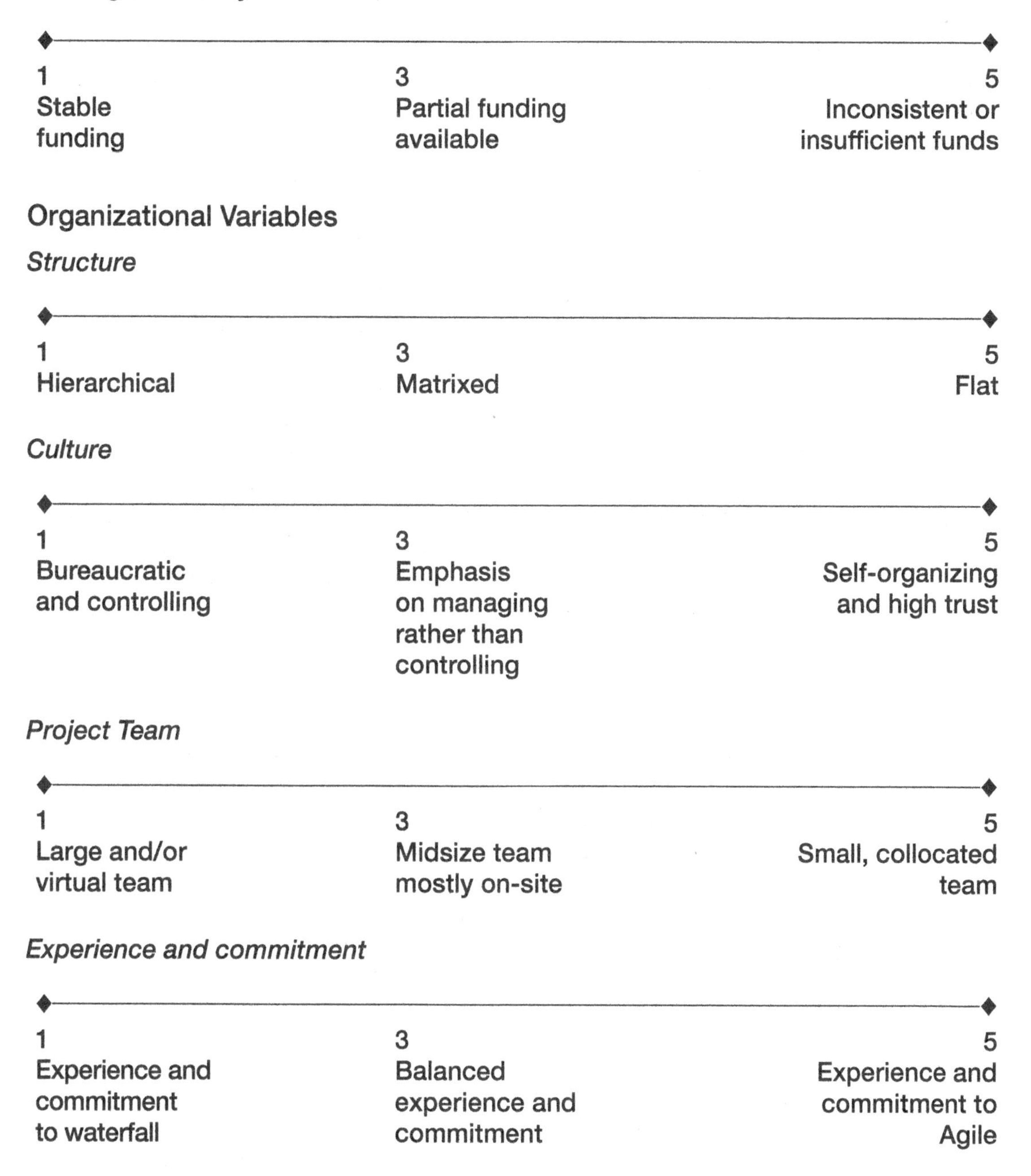

Once you have gone through the process of rating your project or deliverables, the best approach may be very obvious. However, if you want to take your evaluation a bit further, you can create a visual display to help you see the big picture.

CREATING A VISUAL DISPLAY OF THE VARIABLES

A radar diagram is a good way to help you visualize the ratings for all the variables on your project. To create a radar chart, follow these steps:

> **Radar chart:** A graphical chart that displays multiple quantitative variables on numeric axes.

1. Start by rating each variable on a scale of 1–5 (as shown above);
2. In Excel, or some other spreadsheet, enter the variables in a column;
3. Enter the rating for each variable in the next column;
4. Select the cells with the data;
5. Go to the "Insert" tab, and choose the waterfall group of charts; and
6. Select a radar chart.

Figure 2-1 shows a radar chart with three different projects.

You can see how the building construction project is clustered around the center of the chart. All the ratings are 1 or 2. This indicates a waterfall approach is best. The marketing campaign has ratings of 2–4, which means it is a good candidate for a hybrid approach, as some aspects of the project can leverage the stability of waterfall and others the adaptability of Agile approaches. The digital product has ratings of 4–5, which means it will do well with Agile approaches.

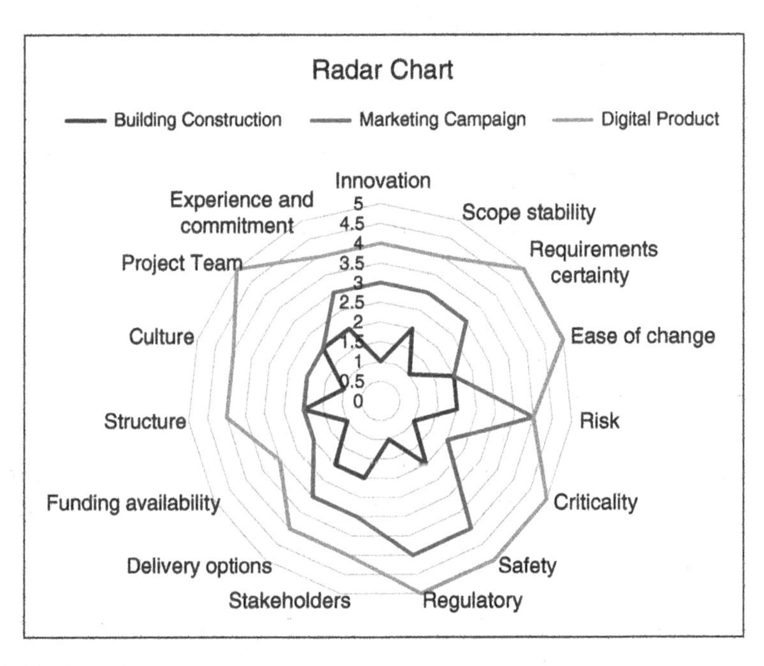

FIGURE 2-1 Radar chart.

SUMMARY

In this chapter we looked at three categories of variables to help evaluate the best development approach for deliverables. Product variables include:

- Innovation;
- Scope stability;
- Requirements certainty;
- Ease of change;
- Risk;
- Criticality;
- Safety; and
- Regulatory.

Project variables include:

- Stakeholders;
- Delivery options; and
- Funding availability.

Organization variables include:

- Structure;
- Culture;
- Project team; and
- Experience and commitment.

We also introduced a way of rating each of the variables on a scale of 1 to 5 and then created a radar chart to get a visual display of a project. Assessing the variables of your project and deliverables lets you leverage the hybrid approach so you can mix and match development approaches, techniques, and methods to find the best way to work for your project.

Key Terms

criticality
radar chart
requirement
risk

3

Project Roles

You can have an awesome schedule and a great risk management plan, but those won't get the project done. It is people who get projects done. In this chapter we will look at roles that are commonly found on projects in a predictive environment and those commonly found in an Agile environment.

Understanding the responsibilities and characteristics of the key project roles will help you tailor project staffing to meet the needs of hybrid projects. We'll review five key project roles:

- Project sponsor;
- Project manager;
- Product owner;
- scrum master; and
- Project team.

Once these five roles are defined, we will look at options for staffing a hybrid project.

PROJECT SPONSOR

Sponsor: The person who provides project resources and supports the project manager in meeting the project objectives.

A project sponsor is a person, usually in a management role, who provides resources and oversight for the project and support for the project manager. In this context, resources include financial resources as well as team resources. In many organizations it is the sponsor who charters a project.

The sponsor's position in the organization is consistent with the size and importance of the project. For example, a project that spans the entire organization, such as an acquisition project, would likely have a sponsor in the C-suite (chief operating officer, chief financial officer, etc.). For a project that only involves one department, it may be the manager or director for that department.

Project sponsors have responsibilities associated with initiating projects, up-front planning, monitoring progress, and supporting the project manager. We'll look at each of these areas of responsibility in more detail.

Initiating Projects

The concept for a project can come from anywhere. Once the project is authorized, usually by a project management office (PMO), a portfolio steering committee, or some other authorizing body, a project sponsor is assigned. Sometimes it is obvious who the sponsor will be, such as the chief information officer when replacing a major system in an organization. In other situations, the sponsor and the customer are the same person, such as when a project is done under contract and the customer who is paying for the project also functions as the project sponsor.

Regardless of how the sponsor is assigned, their responsibilities for initiating a project include:

- **Championing the project at the executive level:** Most organizations have more projects than they have resources to deliver them. The sponsor negotiates for resources at the corporate level, such as team members and budget. They also continue to champion the project and the benefits it will provide throughout the project.
- **Providing financial resources:** In some cases, the sponsor pays for the project out of their department budget; in other cases they will negotiate and gain funding from other sources.
- **Approving the project charter:** A project charter is often used to get a high-level understanding of the project, including objectives, expected benefits, and so forth. The sponsor may work with the project manager to develop the charter or the sponsor may develop the charter and hand it off to the project manager. Their signature signifies approval for the project and provides authorization for the project manager to begin work.

Up-Front Planning

At the start of the project the sponsor and project manager often work together to define some of the high-level information that will guide the development of more in-depth plans. Sponsor responsibilities during the up-front planning for the project include:

- **Providing the initial high-level requirements and information about the project:** Much of the high-level information is included in the project charter. However, there is usually additional information, such as the high-level requirements, assumptions, constraints, variance thresholds, and other information that the sponsor provides during the early stages of planning.

- **Determining the priority among project constraints:** Project managers are constantly balancing scope, schedule, cost, resources, quality, and risk. The project sponsor identifies which are most important so the project manager can make decisions aligned with the relative importance of these constraints.
- **Approving the baselines:** The project sponsor will assess and approve the project schedule and budget baselines. They are a second set of eyes to ensure the schedule and budget reflect the risks involved, resource availability, and other variables. The baseline will be used to measure progress throughout the project.

Monitoring Progress

Once the project is underway, the sponsor is not as involved. They generally stay informed via status reports and are available as needed. Throughout the project, the sponsor may be called on to:

- **Monitor project progress:** The sponsor is the recipient of all project status reports. They will review the schedule status, budget status, projected work for the next reporting period, and any new significant risks and issues.
- **Review all variances outside the acceptable variance threshold:** If a project variance is outside the established threshold, the sponsor will review the situation and likely work with the project manager to determine appropriate actions to bring performance back in line with the baseline.
- **Approve major changes to the project:** Any significant change to the project scope, schedule, and budget must be approved by the sponsor. This usually means additional scope, but it can also include changes to resources, schedule, or budget.

Supporting the Project Manager

The project manager may have a dotted-line reporting responsibility to the sponsor of the project for the duration of the project. Whether or not the project manager reports to the sponsor, the sponsor can support the project manager in the following ways:

- **Resolve conflicts outside the project manager's authority:** As project managers, we may find ourselves in situations where we have people on our team with more position-power than we have. If there are issues with those team members, or if a conflict involves someone outside the immediate project team, the sponsor generally has the position-power, or political clout to remove roadblocks and address issues more effectively than the project manager.
- **Provide mentoring and coaching to the project manager as appropriate:** Whether it is working with difficult stakeholders, navigating political situations, or

employing emotional intelligence, project managers occasionally need someone to provide mentoring or coaching. A sponsor often has the experience and perspective to be able to help the project manager in these situations.

- **Provide insight into the high-level organizational strategy and objectives:** When faced with a project decision, it is useful to have insight into the organization's strategy and objectives to ensure the project is aligned with the direction the organization wants to go. Since project managers often don't have access to this knowledge, they may turn to the sponsor to get guidance and insight to help guide them in the right direction.
- **Manage corporate politics that could affect the project:** Managing a project is challenging enough on its own, but coupled with navigating corporate politics, it can be downright exhausting! The project sponsor can provide political cover for the project manager and keep the politics away from the project.

PROJECT MANAGER

The project manager is accountable for the overall success of the project. This includes demonstrating leadership behaviors while working with people and demonstrating good management practices for achieving outcomes. We'll look at activities for both leadership and management.

Project manager: The person accountable for leading the team to deliver the project outcomes.

Leadership Skills

Leadership skills involve working with the team, engaging with stakeholders, and demonstrating leadership behaviors. Project managers demonstrate leadership in the following ways:

- **Establishing a supportive environment for team members:** People are more likely to give their best efforts if they feel supported and valued. One of the most important jobs for the project manager is to create and maintain an environment that provides psychological safety. This enables team members to do their best work and thrive.
- **Managing stakeholder expectations:** Anyone who has worked with large and/or diverse groups of stakeholders knows that trying to manage stakeholders is a fool's errand. However, we can engage with stakeholders and manage their expectations. Project managers who are successful in working with stakeholders have better outcomes than those who ignore stakeholders or merely tolerate them.
- **Employing leadership skills:** Leadership skills are one of the most important keys to project success. Project managers are constantly drawing on these

skills—for example, communicating effectively, facilitating meetings, solving problems, influencing without authority, and negotiating. Even though we often don't have position authority over our team members, we can and should always demonstrate leadership skills.

- **Demonstrating integrity, stewardship, fairness, and a positive attitude:** Our team looks to us to set the bar on expected and acceptable behavior. Therefore, in addition to the leadership behaviors identified above we must also demonstrate integrity and ethical behavior at all times. One way to do this is to be impartial, fair, and transparent in our decision-making. We should also show respect and demonstrate stewardship for the environment and company property. Finally, we should maintain a positive attitude. Our team members will tend to follow our lead with regards to upbeat energy or a defeatist and whiny attitude. Demonstrating positive energy, especially when confronting challenges and setbacks, sets an example for our team members to follow.

Management Skills

In addition to leadership, the other part of a project manager's role is to manage the project to achieve the desired outcomes. This entails a different set of activities, such as:

- **Developing plans:** A big part of managing predictive projects is developing plans. This is an ongoing process throughout the project. As more information is known, our plans get more detailed and more realistic. When risks or changes occur we need to update our plans. Thus, while a majority of the planning is done toward the start of the project, we aren't done with the plans until the project is done!
- **Establishing project systems:** One way we support our team is by making it easy to accomplish their work. Obviously, where there are processes and systems that have been established by the organization, such as defect reporting, quality assurance, and change management, we need to follow those processes. Where there aren't organization-driven processes, we can establish set ways of working that support the project work. For example, set meeting times, schedules for reporting, common terminology, and so forth.
- **Keeping the project on schedule and within budget:** One of the most prevalent definitions of project success is on time and on budget. Therefore, many of the behaviors listed are meant to enable timely delivering within the approved budget.
- **Managing issues, risks, and variances:** Projects are a breeding ground for uncertainty, risks, and issues. In addition to having processes and systems in place to manage risks and issues, we also need to be skilled in finding ways to prevent them from happening in the first place. Once a risk, issue, or variance has occurred, we

need to work with the team to determine and evaluate corrective actions to get project performance back on track.
- **Collecting project data and reporting progress:** Managing stakeholder expectations and keeping the project on schedule and on budget depend on being able to see and interpret information. Therefore, we need to collect and analyze data and then present it in an effective and compelling way.

PRODUCT OWNER

Product owner: The person accountable for performance of a product.

Product owners typically work in product development, especially with digital and software projects. From a project perspective, they are usually involved with Agile projects; however, many organizations are switching to a product line organizational structure. Therefore, product owners may be involved with projects that use an adaptive approach or a predictive approach. The product owner primarily performs functions associated with the product, though there are some aspects of their work that concern meetings and interactions with the team and stakeholders.

Product Functions

Typical product functions that product owners perform include:

- **Establishing the vision for the product:** At the start of a project, they will develop a vision statement for the project and/or product.
- **Choosing and prioritizing features to maximize value:** Using market research, customer feedback, and other sources of information, the product owner identifies the most important features for the product and prioritizes them for the team.
- **Managing and prioritizing the backlog:** A backlog is a repository of project work, features, and/or requirements. The product owner prioritizes the backlog so the team is always focused on the most valuable work. They can also update the backlog by adding or removing work or reprioritizing existing work.
- **Determining acceptance criteria:** Each deliverable has acceptance criteria that must be met before it can be accepted. It is the product owner who establishes and communicates the acceptance criteria.
- **Accepting or rejecting deliverables:** At the end of an iteration the team demonstrates the work they have done. The product owner accepts the work, requests changes, or does not accept the work, explaining why it is not acceptable.

People Activities

Another aspect of the product owner role is working with the team and various other stakeholders. This includes:

- **Ensuring the team understands the backlog items:** The team will take the information from the backlog and create deliverables. The product owner makes sure that the team understands the information on the backlog and is available to answer questions and provide clarification.
- **Attending planning meetings, demonstrations, and retrospectives:** The product owner is usually in close proximity to the team. They often have daily interaction with the people working on the project. Part of their interaction includes attending planning meetings, demonstrations, and retrospectives.
- **Working with external stakeholders:** Part of the product owner's role is to work as a liaison with external stakeholders. They communicate stakeholder needs and wants to the team and manage stakeholder expectations. This allows the team to focus on development work.
- **Remaining engaged and available to the team:** A primary difference between a sponsor and a product owner is that the product owner is always available to respond to questions and clarify areas of uncertainty. This reduces time spent waiting for a response and rework based on assuming answers to questions.

SCRUM MASTER

Scrum master: The person who supports the team in maintaining alignment with Agile values and principles.

A scrum master is used on projects that work with an Agile methodology. Their role is to support the team in accomplishing the work and following Agile methods. They are a facilitator for the team and support the work of the project.

Facilitation

A scrum master is much more a facilitator than a manager. As a facilitator, scrum masters will:

- **Help teams self-manage and self-govern:** Many Agile teams are self-managing and self-governing. The scrum master supports this by making sure the team has what it needs to self-manage.
- **Facilitate communication between the team and other stakeholders:** scrum masters allow the team to stay focused on the work at hand by engaging

stakeholders and facilitating conversation. This serves to protect the team from outside interference and other distractions.
- **Facilitate meetings:** Agile methods have several types of meetings, such as daily stand-ups (aka scrums), iteration reviews, iteration planning, and retrospectives. The scrum master facilitates these meetings. As part of the retrospective, the scrum master helps the team improve their processes.

Support

In a support role, the scrum master will:

- **Employ servant leadership:** As a servant leader, the scrum master puts the needs of the team first. Scrum masters lead by supporting the team in accomplishing work rather than directing the team.
- **Remove barriers and impediments:** Barriers and impediments delay progress. During the daily stand-up meetings, team members communicate any barriers or impediments (also known as blockers) to getting their work done. The scrum master follows up and endeavors to resolve and remove any barriers and impediments.
- **Provide guidance on Agile methods:** The scrum master ensures the team is following Agile processes and using Agile methods appropriately. This can include coaching, guidance, and education about Agile methods and the benefits they provide.
- **Assist the product owner:** The scrum master works closely with the product owner in keeping the backlog up to date. Scrum masters also assist in communicating the project vision and engaging with stakeholders as appropriate.

With hybrid projects, you may see the term "Agile project manager" or "Agile delivery lead." These roles fulfill the same responsibilities as a scrum master, but they may have additional responsibilities as appropriate in a hybrid environment.

THE TEAM

The team is the group of people performing the project work. On large predictive projects, there may be a subset of the project team called the project management team. This group is composed of team leads or other individuals that help manage aspects of the project. Team members provide the following:

- **Subject matter expertise:** Team members generally have deep knowledge and insight in their areas of specialty. They contribute this expertise in planning, estimating, doing the work, and managing risks.

- **Knowledge for planning the work:** The most effective plans are those that are developed collaboratively with the project manager or scrum master and the team members. As subject matter experts, team members provide valuable information for what needs to be done and how it can best be accomplished.
- **Estimates for the work:** Team member expertise and experience is necessary to develop reliable estimates for effort and duration. Team members can also provide information on the likely cost of materials and equipment as well as contractor rates.
- **Work on activities as assigned:** Throughout the project, team members apply their knowledge and skills to accomplish project activities.
- **Information on risks and issues:** Team members know from experience what can go wrong on a project. As such, they can help identify and suggest responses for risks and issues that arise.
- **Participation in team meetings:** On predictive projects there are usually weekly team meetings. On Agile projects there are daily stand-up meetings, planning meetings, demonstrations, and retrospectives.
- **Leadership behaviors throughout the project:** All team members, regardless of their position on the team, should demonstrate leadership behaviors. This includes active participation in problem solving, decision-making, and brainstorming. Team members should demonstrate collaboration, transparency, respectful communication, and other leadership behaviors.

Generalizing Specialists

I-shaped people: People who have deep knowledge in one specific area.

T-shaped people: People who have deep knowledge in one area and broad knowledge or skills in complementary areas.

There is a concept in Agile called generalizing specialists. This refers to people who have deep knowledge in one area and broad knowledge in other areas. These people are also known as T-shaped people, where their area of expertise is the vertical line of a T and the broad areas of expertise are represented by the crossbar, or horizontal line, as shown in Figure 3-1.

The benefits of T-shaped team members are that they can represent multiple perspectives and perform activities in a number of areas. This reduces the need for knowledge transfer and handoffs among team members, thereby saving time. They can also help resolve bottlenecks as they can step in to help in a number of areas.

Figure 3-2 shows an I-shaped person. An I-shaped person is the opposite of a T-shaped person. I-shaped people only have deep knowledge in one area and little to no knowledge in other areas.

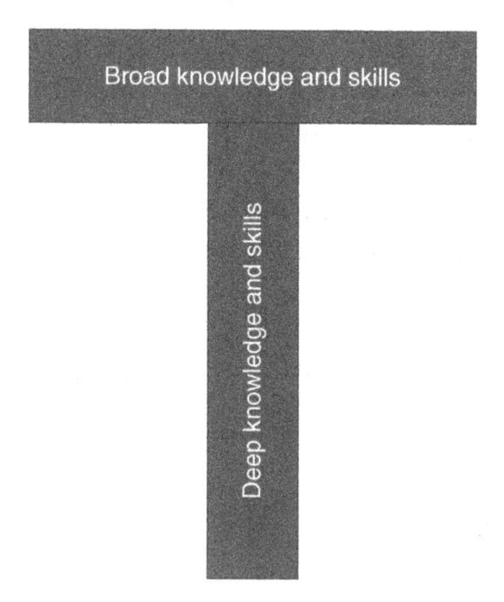

FIGURE 3-1 T-shaped people.

FIGURE 3-2 I-shaped people.

When building a team, it is useful to have as many generalizing specialists as possible to keep your project running smoothly.

HYBRID OPTIONS

On a hybrid project you can mix, match, and combine roles to meet the needs of your project. Here are a few examples.

- The sponsor may also fulfill a product owner role.
- There may be a sponsor for the overall project and a product owner for one part of the project. The sponsor would charter the project and provide information about

the overarching strategic direction and expected benefits. The product owner would use that information to make decisions about a specific deliverable for the project.
- A project manager may be accountable for the overall project but work with a scrum master for a software deliverable. The project manager may keep the schedule and budget for the project, keeping placeholders in the schedule for iterations and releases. Meanwhile the scrum master would facilitate the development team using iterations and releases.
- A project manager may adopt methods used in Agile projects while using a predictive framework for leading the team and managing the project. They may implement daily stand-ups, employ servant leadership, use task boards, and conduct retrospectives.

For hybrid projects, it's less about the title and more about what makes sense for the project.

SUMMARY

In this chapter we looked at five project roles: the sponsor, project manager, product owner, scrum master, and project team. Sponsors have responsibilities for initiating, up-front planning, monitoring progress, and supporting the project manager. The project manager performs both management and leadership activities. Product owners have product responsibilities and people responsibilities. Scrum masters focus on facilitation activities and support activities for the team. Team members provide skills and knowledge in their areas of expertise. Some team members are T-shaped, with deep and general knowledge, and others are I-shaped, with deep knowledge in one area of expertise.

Key Terms

generalizing specialists
I-shaped people
product owner
project manager
scrum master
sponsor
T-shaped people

4

Launching a Hybrid Project

The very beginning of a project is always exciting. We're thinking about the outcomes and deliverables we will produce, the team and other stakeholders we will work with, and the difference the project will make. Depending on the type of project you are working on, you may launch a project with a project charter, a project vision statement, or some combination of the two. Both these documents are high-level descriptions of the project, but the amount of detail and the structure are quite different.

When you start a project, you don't have all the information and facts you need; however, you still need to develop plans. In the absence of facts, project managers make and document assumptions using an assumption log. Constraints can limit options on a project, and those are also documented in the assumption log. The assumption log is a dynamic document that is updated throughout the project.

In this chapter we'll look at the purpose and structure of both a charter and a vision statement. Then we'll look at a vision statement and charter for a case study we will use to demonstrate concepts throughout the rest of the book.

VISION STATEMENTS

A vision statement is a brief and compelling description of the desired future state. Organizations have vision statements, and projects can have them as well.

Organizations' Vision Statements

Let's start by looking at a few examples of some organizations' vision statements.

Oxfam: A just world without poverty.

Apple: Make the best products on earth and leave the world better than we found it.

LinkedIn: Create economic opportunity for every member of the global workforce.

Sony: To be a company that inspires and fulfills your curiosity.

What do each of these vision statements have in common? They meet the following four criteria for good vision statements.

Brief: A vision statement should be no more than one or two sentences. Instagram's original vision statement was "Capture and share the world's moments." Nike's original vision statement was "Crush Adidas." Both of these are brief, compelling, and memorable.

Compelling: A vision statement should motivate people to want to go to work. It should draw people in. Oxfam's vision statement gives you a good reason to get up and go to work in the morning. I don't know anyone who doesn't want a just world without poverty!

Memorable: Everyone at the organization should be able to recite the vision statement. For that to happen, they have to be both brief and compelling. Cold Stone Creamery's vision statement is "The ultimate ice cream experience." That's a good vision statement because it is definitely more memorable than something like "We make good ice cream."

Begin with the end in mind: Vision statements should speak as if the future the organization is striving for is here now. This can be written as a problem that has been solved or changes that have been made. The vision statement for the Alzheimer's Association is "A world without Alzheimer's." That is not the state we are in now, but it is brief, compelling, and memorable. It speaks of a future state as if it is currently happening.

PROJECT VISION STATEMENTS

Project vision statement: A brief and compelling statement that describes the future state after the project is complete.

Like organizational vision statements, project vision statements should be brief, compelling, memorable, and with the end in mind. When Boeing launched the 777, they wanted a mid-sized aircraft that was technologically advanced and could operate in extreme environments (such as Denver, which is a mile above sea level), make overseas flights, and tolerate temperature extremes. All those are requirements, but none of them is particularly compelling as a vision statement. However, stating it as "Denver to Honolulu on a hot day" creates a visual image in the mind's eye and is compelling and more memorable than a list of requirements.

Another compelling vision statement was when John F. Kennedy stated ". . .before the end of the decade, landing a man on the moon and returning him safely to the Earth." For the men and women working on that project, the statement was motivating, memorable, and inspiring. That vision statement wasn't compelling just to the program staff but to most of the people on the planet as well—and it started with the end in mind.

How can this translate to our projects? Most of us aren't working on projects that are as riveting as the Apollo program; however, our projects can still benefit by having a good vision statement. In *Crossing the Chasm*, Geoffrey Moore developed a recipe for a vision statement for new products that goes like this:

For _____ *identifies the customer*_____
Who ____ *provides more detail*_________
The ___ *names the product*________________ is a ____ *describes the product*______
That ___ *identifies what the product does*______
Unlike ___ *describes competitor products* ______
Our product __ *provides differentiating factors* ____.

An example of a vision statement for a new adult education center that intends to provide upskilling and reskilling training could be:

For adults
Who are changing careers
The Career Upskilling Center is a supportive place to learn new skills
That prepares students for in-demand jobs.
Unlike degree programs
The Career Upskilling Center provides only the skills needed when they are needed.

While this is not as brief and compelling as "Crush Adidas," it does provide a simple way to identify the customers, the product, and how the product will meet their needs. It also identifies differentiating factors. This is a good elevator pitch for the project, and it gives the team and key stakeholders a big-picture vision of what the project will accomplish.

Vision statements are often used on new product development projects.

PROJECT CHARTER

Project charter: A document that formally authorizes a project and provides a high-level description of the project.

A project charter has quite a bit more detail than a project vision statement. A charter is usually developed either by the project sponsor or in a joint effort between the project sponsor and the project manager. It provides high-level information and functions as an authorization to begin the project.

The contents of the charter should be tailored to meet the needs of the product, project, and organization. Table 4-1 describes elements that are commonly found in a charter.

TABLE 4-1 Project Charter Elements*

Element	Description
Project purpose or justification	The reason the project is being undertaken. May refer to a business case, the organization's strategic plan, external factors, a contract agreement, or any other reason for performing the project.
High-level project description	A summary-level description of the project. May include information on high-level product and project deliverables as well as the approach to the project.
High-level requirements	The high-level conditions or capabilities that must be met to satisfy the purpose of the project. Describe the product features and functions that must be present to meet stakeholders' needs and expectations. This section does not describe the detailed requirements, as those are covered in requirements documentation.

*From *The Project Manager's Book of Forms*, 3rd edition, by Cynthia Snyder.

TABLE 4-1 Project Charter Elements (*Continued*)

Element	Description
Project objectives and success criteria	Project objectives are usually established for at least scope, schedule, and cost. Scope objectives describe the scope needed to achieve the planned benefits of the project. Time objectives describe the goals for the timely completion of the project. Cost objectives describe the goals for project expenditures.
High-level risks	The initial risks at a project level, such as funding availability, new technology, or lack of resources.
Summary milestone schedule	Significant events in the project. Examples include the completion of key deliverables, the beginning or completion of a project phase, or product acceptance.
Summary budget	The estimated range of expenditures for the project.
Key stakeholders	A list of significant people who have an interest and an influence on the project success. Stakeholders who do not require significant engagement may be noted in a stakeholder register rather than the charter.
Project approval requirements	The criteria that must be met for the project to be accepted by the customer or sponsor.
Assigned project manager, responsibility, and authority level	The authority of the project manager with regard to staffing, budget management and variance, technical decisions, and conflict resolution. Examples of staffing authority include the power to hire, fire, discipline, and accept or not accept project staff. Budget management refers to the ability of the project manager to commit, manage, and control project funds. Variance refers to the variance levels that require escalation for approval or re-baselining. Technical decisions define or limit the authority of the project manager to make technical decisions about deliverables or the project approach. Conflict resolution defines the degree to which the project manager can resolve conflict within the team, within the organization, and with external stakeholders.
Name and authority of the sponsor or other person(s) authorizing the project charter	The name, position, and authority of the person who oversees the project manager for the purposes of the project. Common types of authority include the ability to approve changes, determine acceptable variance limits, impact inter-project conflicts, and champion the project at a senior management level.

In general, charters are used for predictive projects, and they don't change unless there is a significant change in a project objective.

CASE STUDY

Throughout this book we'll be looking at a case study for the Greek God Wine Group to demonstrate predictive, adaptive, and hybrid options for working on a project.

Background

The Greek God Wine Group (GGWG) is a successful winery. Due to its success, GGWG is expanding its operations by purchasing an older winery in Central California. The winery will be called Dionysus Winery. The winery has 250 acres of producing vineyards and a large but aging production and storage facility.

GGWG plans on renovating the winery and constructing a boutique hotel, restaurant, and tasting room. To ensure efficient operations, they will be installing a winery management system for inventory, production, and vineyard management. They have big plans for the winery including a wine club and many social and recreational events.

Tony Dakota, a project manager with experience in both predictive and adaptive management methods, has been brought in to manage the project.

Tessa Barry is the chief operating officer for GGWG. She will be the sponsor and is accountable to GGWG's board of directors for the successful launch and operations of the new winery.

Case Study Vision Statement

At the start of the project, Tessa and Tony met to develop a vision statement for the winery. They came up with the following:

For adults
Who appreciate wine and good times
The Dionysus Winery is a state-of-the-art winery
That caters to the fun and frivolity in all of us.
Unlike other wineries that only offer wines, clubs, and tasting rooms
The Dionysus Winery will offer education and recreation activities, unique lodging, and innovative, locally sourced gourmet food.

Case Study Charter

Keeping the vision statement as a guide, Tony and Tessa developed the following Charter.

Dionysus Winery Charter

PROJECT PURPOSE

Provide a fun, upscale, unique winery experience for individuals, couples, and groups.

HIGH-LEVEL PROJECT DESCRIPTION

This project includes all work necessary to:

1. Renovate the existing wine production facility;
2. Renovate the existing wine storage facility;
3. Develop a new winery management system for vineyard, inventory, and production management;
4. Construct a boutique hotel with 50 rooms;
5. Construct a restaurant with a 200-person capacity;
6. Construct a tasting room with room for 25 tasting stations and 250 guests;
7. Establish a wine club;
8. Hold a grand opening event to showcase Dionysus wines and social activities;
9. Hire and train all staff;
10. Communicate with all stakeholders as appropriate;
11. Lead multiple teams to deliver the project;
12. Manage scope, schedule, budget, resources, and risks, throughout the project; and
13. Provide status updates to GGWG on a monthly basis.

HIGH-LEVEL REQUIREMENTS

The project shall meet the following conditions and capabilities:

1. Utilize local labor wherever possible;
2. Use a hybrid approach, selecting the best delivery method for each deliverable while maintaining a master plan, schedule, and budget;
3. Employ eco-friendly materials and practices;
4. Maintain compliance with all state and local laws, regulations, and ordinances regarding vineyards, wineries, tasting rooms, lodging, staffing, and activities;
5. Build the hotel to 4-star AAA rating standard;
6. Build and staff the restaurant to 4-star AAA rating standard; and
7. The grand opening shall include Dionysus wines, food from the restaurant, music, and fun activities.

PROJECT OBJECTIVES AND SUCCESS CRITERIA

The project will be considered a success if the following objectives are met:

- Manage the scope for the production, storage, vineyard, and all construction deliverables;
- Evolve the scope for the vineyard, inventory, and production system;
- Identify and evolve the scope for the grand opening and the initial social and recreational activities;
- Hold the grand opening no later than 12 months from the project start date; and
- Maintain budget performance within +/− 10%.

HIGH-LEVEL RISKS

Areas of greatest uncertainty for this project are:

- Availability of staff with needed skills;
- Timeliness of permits and certificates of occupancy;
- Availability of materials; and
- Cost of materials.

SUMMARY MILESTONE SCHEDULE

The following milestones will be used to track progress.

- Start up;
- Renovation complete (production and storage facilities);
- Construction complete (hotel, restaurant, tasting room);
- Winery management system deployed;
- Staffing and training complete;
- Wine club established;
- Grand opening; and
- Transition to operations.

SUMMARY BUDGET

The project budget is $9,000,000. This does not include contingency or management reserve.

KEY STAKEHOLDERS

The initial stakeholders are identified as:

- Greek God Winery Group;
- Sponsor;
- Team;
- Vineyard staff;
- Local entertainers and activity vendors;
- Other wineries;

- Local residents;
- Tourists/visitors; and
- Media.

PROJECT APPROVAL REQUIREMENTS

The sponsor will consider the project complete when the following criteria are met:

1. All requirements fulfilled;
2. All permits and licenses are in hand;
3. All venues are fully staffed, and staff is trained;
4. The grand opening is complete; and
5. All operations have been transitioned to staff.

ASSIGNED PROJECT MANAGER, RESPONSIBILITY, AND AUTHORITY LEVEL

The authority of the project manager, Tony Dakota, for staffing, budget management, and technical decisions, are as follows:

- The authority to manage all team leads, who will in turn manage and lead their team members; and
- The authority and accountability to estimate, monitor, and control all funds, other than salaries and operating expenses.

The project manager does not have the authority to:

- Hire or fire staff, but his input will be strongly considered;;
- Sign contracts; or
- Make technical decisions regarding viniculture (science of wine grapes) and enculture (science of winemaking).

NAME AND AUTHORITY OF SPONSOR

The sponsor for this project is the chief operating officer of the Greek God Wine Group, Tessa Barry. Tessa has overall accountability for the success of the new Dionysus Winery. This includes approving schedules, budgets, staffing, changes, and variances. Tessa is the escalation point for risks, issues, and conflicts that are outside the project manager's authority.

Tessa is the champion the project at a senior management level.

Signatures:

Anthony Dakota	*Tessa J. Barry*
Project Manager	Project Sponsor

One of the keys to a successful project is starting out right. Having a compelling vision and a project charter are good first steps in delivering a successful project.

ASSUMPTIONS AND CONSTRAINTS

Assumption: Something that is considered true for purposes of the project.

Constraint: A limiting or restricting factor.

Assumption log: A dynamic document that is used to identify and track assumptions and constraints.

From the very start of the project, the project manager is making assumptions and working with constraints. For example, when starting a project, there may be an assumption that project staff will have sufficient skills to accomplish the work. At the beginning of the project, you may not know who will be working on the project, but you can't build a schedule or budget without making some assumptions. If at some later date you find that an assumption is not true, you may need to update the schedule or budget to reflect the new information. It's important to remember that an assumption that isn't true may lead to a risk for the project.

Projects operate under a set of constraints. Constraints can be articulated via requirements, regulations, and policies. Examples of these type of constraints include maintaining compliance with local ordinances or following company policy with regard to contracts. Other constraints are more project specific, such as fixed delivery dates or funding limitations.

It is a good practice to document both assumptions and constraints in an assumption log. The following are common contents in an assumption log:

- **Category:** Having categories of assumptions and constraints, such as resources, schedule, budget, and so forth, helps organize and group assumptions and constraints.
- **Assumption/constraint:** A brief description of the assumption or constraint with enough information to understand the assumption or constraint but without exhaustive detail.
- **Status:** The status of the assumptions, such as pending, active, or closed.
- **Comments:** Any additional useful information. This can include historical information, current activities to validate the assumption, implications if the assumption is not valid, and so forth.

The assumption log may also include fields for a responsible party to validate the assumption, a due date, and actions.

Table 4-2 shows an example of the start of an assumption log for the Dionysus Winery project.

TABLE 4-2 Dionysus Winery Assumption Log

ID	Category	Assumption/Constraint	Status	Comments
A1	Skills	We assume there is a sufficient local talent pool to staff the vineyard, winery, and tasting room.	Pending	Since there are several wineries in the area, there are likely people with the needed skills.
A2	Operations	We assume the vineyard is in good shape and will be managed by existing Greek God Wine Group staff.	Confirmed	Vineyard operations are not part of this project, but the staff are considered important stakeholders for the project.
A2	Resources	We assume construction resources are available at market rates.	Pending	There has been a shortage of building materials. We assume the shortage will be addressed and not affect this project.
C1	Regulatory	The production, storage, and winery facilities must be updated to the latest codes per the Alcohol Beverage Control Board (ABC).	Active	New regulations have been passed, and the production and storage facilities are out of compliance.

Given the dynamic nature of projects, we don't always have facts about how things will evolve. Therefore, it is important to make and document assumptions about our project. Your assumption log is a living document that will continue to evolve and be updated throughout the project.

SUMMARY

In this chapter we looked at two documents that can be used at project start-up, a project vision statement and a project charter. We also described how an assumption log can be used to document project assumptions and constraints.

We introduced the Dionysus Winery case study that will be used throughout the book. As part of introducing the case study, we documented a vision statement, project charter, and assumption log for the Dionysus Winery.

Key Terms

assumption
assumption log
constraint
project charter
project vision statement

5

Hybrid Project Planning and Structure

Hybrid projects require up-front thought and planning to structure them appropriately. By structure I mean determining the development approach you will use for each deliverable, the life cycle phases, and how you will integrate adaptive and predictive approaches in your project. You will also need to consider which elements should be in your project management plan and how detailed they need to be. Deliverables usually have various reviews and sign-offs. You can record this information plus milestones and key delivery dates in a project roadmap.

In this chapter we will cover fundamental concepts in project planning. These concepts apply throughout the project, regardless of the development approach you are using. We will review common elements in a project management plan and discuss how to tailor them. Then we we'll describe a project life cycle and project phases. We'll demonstrate these concepts by showing how this information would be applied to the Dionysus Winery project. Last, we'll introduce a project roadmap and show how it provides a high-level structure for the project.

PLANNING FUNDAMENTALS

There are two fundamental truths in project management:

1. You can't know everything at the start; and
2. You have to balance competing demands.

We'll start by looking at a planning technique that accounts for the uncertain nature of projects.

Progressive Elaboration and Rolling Wave Planning

Once you have a project vision statement and/or a project charter, you can start planning the project. There is usually some degree of planning at the start of the project and then varying degrees of planning as the project progresses.

As discussed in Chapter 2: "Choosing a Development Approach," projects with evolving scope don't need a lot of up-front planning. With this type of project there isn't a lot of certainty about the scope, and therefore it doesn't make sense to spend time estimating resources, duration, and budget in detail at the start of the project.

For projects with well understood and well-defined scope, there is more planning up front, but not everything can or should be planned out in detail at the start of the project. This is because while the resources, schedule, and cost estimates may be fairly certain for the first 90 days, once you get too much past that, things become more uncertain. Risks and issues can occur, stakeholders can change their minds, there could be schedule delays, changes in team members, and so on. Therefore, even with waterfall projects, we expect to continually provide more detail to plans and estimates as a project progresses. This concept is known as "progressive elaboration."

> **Progressive elaboration:** Developing more detail as more information is known about the project.

Progressive elaboration is a fact of life on projects. Even though we would love to be able to accurately predict exactly what will be happening on day 210 of our project and how much of the budget we will have spent, it is just not possible.

One way to balance the uncertainty associated with projects and the desire for predictability is to use rolling wave planning. Rolling wave planning ensures that the time spent planning is appropriate for the point of time in the project. Some people use a 90-day horizon—meaning that information for the next 90 days is documented in detail. This can include schedules with

> **Rolling wave planning:** A form of progressive elaboration where activities in the near future are planned in detail, and activities further out in the future are less specific.

activities, dependencies, and defined durations. The cost estimates have a tighter range, and resources are committed. Information from 90–180 days is planned in less detail. Schedules may have less detail, and the cost estimates may have a wider range. The necessary skills may be identified, but some of the activities may not have specific individuals assigned yet. Work that will take place more than 6 months in the future is held at a milestone level with rough estimates for costs.

Depending on the project, you may have shorter planning "waves." For projects with fixed scope that are shorter than 90 days, you can usually do a reasonably complete job of planning at the start and merely fine-tune your estimates along the way.

Progressive elaboration doesn't mean you ignore the milestones and summary level budgets from the charter. It means you don't spend a lot of time trying to get into the low level of detail for the entire project at the start. You elaborate and refine the detail as you progress through the project.

Competing Demands

There's a project management saying: "You can have it good, you can have it fast, or you can have it cheap. Choose two." Said another way, "Scope, schedule, or cost—which is most important?" Scope, schedule, and cost have been referred to as the "iron triangle." While that is a powerful mental image, it doesn't tell the complete story.

To be certain, those are three important constraints in projects, but they aren't the only constraints we need to be aware of. Let's look at a few examples using a project to build a custom home.

Scope and quality are different: Let's say the customer wants tile floors. You can go to the local hardware store and pick up tile that costs $5 for each 12 by 12 tile. But that $5 is a very different quality from an Italian marble that is $20 for the same size tile. The scope of having tile floors may be met, but it may not meet the quality the customer was hoping for. In this example there needs to be a balance between the constraints of quality and budget.

Cost and duration are dependent on resources: In a perfect world team members with the skills you need are readily available, and all the materials you need are local and reasonably priced. We all know that while this is possible, it is not guaranteed. Sometimes team members are on other projects, and sometimes there are supply chain issues that delay schedules and increase prices. This demonstrates the need to balance resources with schedule, budget, and perhaps even scope and quality.

Risks can affect any aspect of the project: Even if you have all the resources you need, a realistic schedule, agreed-upon quality requirements, and sufficient budget, there is no guarantee that a risk, issue, or problem won't come along and negate everything.

Therefore, I consider that every project is subject to at least the following six constraints:

- Scope;
- Schedule;
- Cost;
- Quality;
- Resource; and
- Risk.

In addition, most projects are subject to regulatory constraints that are specific to an industry or profession. And let's not forget stakeholders. Stakeholders are the reason projects exist. Therefore, throughout the project when we are balancing our competing demands, we should also consider stakeholder satisfaction among our constraints.

Thus, beginning with planning a project and continuing throughout, we are always balancing and rebalancing competing demands. This takes constant attention and good judgment in order to deliver your project successfully.

THE PROJECT MANAGEMENT PLAN

A governing document for waterfall projects is the project management plan (aka project plan). The information in the project management plan and the level of detail is dependent on the project and organizational guidance. Developing the project management plan can be a time-consuming and rigorous process for large complex projects. However, the process of developing the plan surfaces valuable information and can help the project run smoothly.

> **Project management plan:** A plan that describes how the project will be planned and managed.

A project management plan usually has a description of the life cycle and phases and several subsidiary plans for key elements of the project. It also describes the necessary key reviews.

Subsidiary Plans

Subsidiary plans address how specific areas of the project will be planned and managed. For example, many projects have a risk management plan that outlines how the team will identify, analyze, and respond to risks.

> **Subsidiary plan:** A component of the project management plan that describes how a specific aspect of a project will be planned and managed.

Table 5-1 provides a description of common subsidiary plans and their contents.

TABLE 5-1 Subsidiary Plans

Plan	Content
Scope management plan	Guidance for how scope will be defined, documented, verified, managed, and controlled; Instructions for gaining formal acceptance and verification of project deliverables; Process for controlling changes to the project scope.
Requirements management plan	Describes how requirements will be collected, tracked, and reported; Describes the configuration management plan for requirements; Defines requirements prioritization criteria; Establishes a requirements traceability structure.
Schedule management plan	Describes the scheduling methodology; Identifies the scheduling tool; Sets the format for developing the project schedule; Establishes criteria for controlling the project schedule.
Cost management plan	Defines the precision level of estimates; Defines the units of measure; Establishes cost control thresholds.
Quality management plan	Identifies quality management roles and responsibilities; Defines the quality assurance approach; Defines the quality control approach; Defines the quality improvement approach.
Staffing plan	Outlines roles, authority, and responsibility; Sets the project organizational structure; Describes staff acquisition and release processes; Identifies training needs.
Resource management plan	Identifies how physical resources will be estimated; Defines how and when resources should be acquired; Outlines information for resource logistics (delivery, storage, management).
Communications management plan	Describes the type of information that will be distributed: level of detail, format, content, and so on; Identifies the audiences for each communication; Outlines the timing and frequency of distribution; Documents a glossary of terms.
Risk management plan	Outlines roles and responsibilities for risk management; Identifies budgeting and timing for risk management activities; Describes risk categories; Provides definitions of probability and impact; Provides a probability and impact (PxI) matrix; Describes quantitative reporting methods if used.

TABLE 5-1 Subsidiary Plans (*Continued*)

Plan	Content
Procurement management plan	Outlines procurement authority, roles, and responsibilities; Documents the standard procurement documents; Identifies contract types; Documents selection criteria.
Stakeholder engagement plan	Identifies the current and desired level of stakeholder engagement for categories of stakeholders; Describes actions to engage stakeholders effectively; Identifies relationships and interdependencies among stakeholders.
Change management plan	Identifies how change requests will be submitted; Documents process for evaluating change requests; Defines the membership of the change control board; Outlines the authority for approving changes.

Of course, there are other plans that may be appropriate for projects, such as a logistics plan, configuration management plan, and safety plan. For some projects more detail than what is outlined above is appropriate; for other projects less information is appropriate.

Tip. Look for ways to combine or reduce the number of subsidiary plans. For example, you may be able to combine the scope and quality plan. Another example is you may not need a procurement management plan if you don't have large purchases.

Tailoring the Project Management Plan for Hybrid Projects

Leading hybrid projects requires you to tailor the project management plan even more than for a waterfall project. Below are some examples of how you might need to tailor components of your project management plan.

- **The scope management plan** may have information on which deliverables will use predictive methods for scope development and which will use more evolutionary methods.
- **The schedule management plan** may indicate that certain deliverables will be scheduled using a Gantt chart and others with a task board. The schedule management plan will also describe how the task board work will be integrated and shown on the overall project schedule.
- **The staffing plan** may address different roles and responsibilities for the team, such as the scrum master, project manager, sponsor, and product owner. There

are different certifications for roles in Agile projects than there are for waterfall projects, so any certification or training requirements would be documented in the staffing plan.

- **The communications plan** generally documents various types of meetings used to communicate status. The types of meetings used on adaptive projects are different than those in a predictive project. Therefore, the expectations for which types of meetings, the frequency, and who is expected to attend would be documented in a communications plan.
- **The change management plan** may identify which deliverables need to go through a change management process and which are considered evolving.

These are just a few examples of how you might tailor a project management plan to meet the needs of a hybrid project.

PROJECT LIFE CYCLES

> **Project life cycle:** A series of phases a project goes through from inception to completion.

A project life cycle is a series of phases that a project goes through from inception to completion. Project phases are specific to the type of work being done. For example, in Chapter 1 we showed a sample life cycle for a construction project as shown in Figure 5-1:

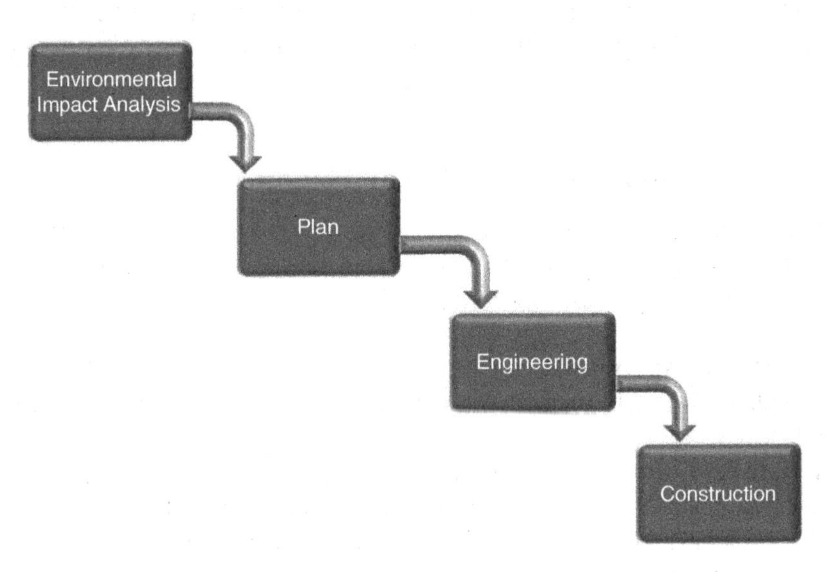

FIGURE 5-1 Construction project sample life cycle.

Obviously, you would not use the same series of phases for a project to develop a research and development (R&D) project. For R&D a more appropriate life cycle might be similar to the one shown in Figure 5-2.

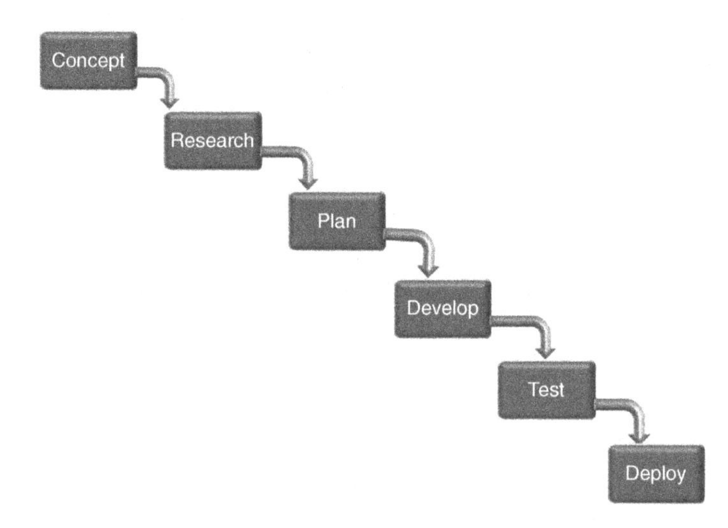

FIGURE 5-2 R&D project sample life cycle.

In this sample life cycle for an R&D project, if the team is using an adaptive development approach, the plan, develop, and test phases would likely be happening iteratively and at the same time. In a waterfall approach the phases would happen sequentially, or possibly overlap. Therefore, the nature of the product influences the development approach, and the development approach influences the life cycle.

Project phases represent the type of work being done. I think it is useful to have a brief description of the work that will be done in each phase and the criteria for advancing from one phase to another. Table 5-2 shows an example of the life cycle I used for writing this book. In addition to listing the phases, it describes the type of work performed in each phase and the criteria for advancing to the next phase.

TABLE 5-2 Life Cycle Phases for Publishing a Book

Phase	Work	Advancement Criteria
Proposal	• Define the concept. • Identify the approach. • Fill out book proposal. • Market research. • Contract.	• Market research is favorable for the proposed concept. • A contract is signed by both parties. • ISBN number is acquired.

(Continued)

TABLE 5-2 Life Cycle Phases for Publishing a Book (*Continued*)

Phase	Work	Advancement Criteria
Write	• Develop detailed outline. • Draft each chapter. • Develop graphics. • Create glossary. • Proofread. • Submit.	• The manuscript and all graphics are submitted to the publisher in compliance with editorial guidelines. • All contractual milestones are met.
Edit	• Content is reviewed for grammar and clarity. • Graphics are reviewed for clarity. • Author review.	• Technical and project editors approve content. • Art department approves graphics. • Author has reviewed all edits and approved final content.
Layout	• Publisher converts all content for print. • Pagination is set. • Graphics are converted to print-ready format. • Cover-design is finalized.	• All content is in print-ready format.
Print	• Pages, are produced and reviewed for accuracy. • Pages are typeset. • Proofs are created. • Plates go to press. • Sheets are printed. • Sheets go to bindery.	• Pages pass QA. • Proofs pass QA. • Pages are printed correctly. • The book is bound and ready for market.
Market	• Book added to catalogs and online resources. • Sales and marketing promote book. • Book is stocked in online and brick and mortar stores.	• Project is complete. • Book moves into operational management.

You can see by reviewing the table above that the phase descriptions are brief. They provide just enough information to indicate the activities happening in each phase and what needs to happen to move into the next phase. In many projects there may be overlapping phases. For example, there may be some overlap in the edit and layout phases and in the print and market phases. In that case you may have phase completion criteria rather than advancement criteria.

A hybrid project that has different development approaches may have a life cycle with phases that run in parallel. There could be a start-up phase for the overall project and then multiple phases where development takes place depending on the type of work. For example, you might see phases for requirements → develop → test for a software deliverable running in tandem with phases for engineering → build → finish work for a construction deliverable. This type of life cycle adds a level of complexity because of the need to coordinate different types of work and deliverables across the project.

KEY REVIEWS

Projects often have reviews at key points in the project. They may occur at the start or end of a phase or when a critical deliverable is complete. The purpose of a review is for relevant stakeholders to review the work, ask questions, see a demonstration (if possible), and ensure that certain criteria (such as advancement criteria or acceptance criteria) have been met.

Examples of key reviews include an integrated baseline review, preliminary design review, and engineering peer review. The descriptions below provide a general idea of what to expect with key reviews.

Integrated baseline review (IBR): An IBR is used to confirm that the project is planned well. The review assesses at least the following:

- A robust risk assessment has been conducted, and risks have been addressed;
- The scope, cost, and schedule baselines are in alignment;
- There are appropriate control systems in place;
- The project management plan is complete and sufficient to move forward.

Preliminary design review (PDR): A PDR confirms that the project can move forward with detailed planning. This review assesses at least the following:

- The preliminary design meets all the requirements;
- The project is able to meet the cost and schedule baselines;
- Technical risks are acceptable and have adequate responses in place;
- All interfaces have been identified;
- Verification methods have been identified.

Technical peer review: A technical peer review is not attended by management; rather, it is an opportunity for peers to review the work. It is generally used in engineering projects. These reviews are often held in the development phase of a project and may be used to review product components or completed products.

As with everything, reviews depend on the nature of the deliverables and the project. Small projects may have a single review before transitioning to operations. Large projects may have as many as 7–10 reviews that are multiday affairs with clients, contractors, and subcontractors.

Phase gate: A review at the end of a phase to determine if the project is ready to advance to the next phase.

A review at the end of a phase is often called a phase gate. A phase gate is used to determine if a project is ready to advance to the next phase. Usually this means that the completion criteria for a phase have been met. Sometimes a project will advance, even if there are one or two items that are still in progress or need to be revised. In rare occasions a project will be terminated at a phase gate because the need for the project no longer exists, because it is clear the project can't meet the objectives, or other reasons.

PROJECT MANAGEMENT PLAN FOR A HYBRID PROJECT

We'll use the Dionysus Winery project to demonstrate how to tailor the components in the project management plan to meet the needs of the project. We'll start with documenting the development approach and life cycle, then discuss which subsidiary management plans would be useful, and finish up with key reviews.

Development Approach

Different deliverables will have different approaches. The construction and renovation will do well with a predictive approach, while the management system can be developed using an adaptive approach.

Tony has met with the project team, and they have documented the development approach for each deliverable along with a brief explanation on why the approach will be used.

TABLE 5-3 Development Approaches for Dionysus Winery Project

Deliverable	Development approach	Comments
Boutique hotel	Waterfall	The requirements for the hotel are well understood and are not likely to change. Much of the planning can take place at the start, and the plans can be baselined and followed.
Restaurant	Waterfall	The requirements for the restaurant are well understood and are not likely to change. Much of the planning can take place at the start, and the plans can be baselined and followed.

TABLE 5-3 Development Approaches for Dionysus Winery Project (*Continued*)

Deliverable	Development approach	Comments
Tasting room	Waterfall	The requirements for the tasting room are well understood and are not likely to change. Much of the planning can take place at the start, and the plans can be baselined and followed.
Renovated wine production facility	Waterfall	The requirements for the wine production facility are well understood and are not likely to change. Much of the planning can take place at the start, and the plans can be baselined and followed.
Renovated wine storage facility	Waterfall	The requirements for the wine storage facility are well understood and are not likely to change. Much of the planning can take place at the start, and the plans can be baselined and followed.
Winery management system	Agile	The management system can start with a list of prioritized features and requirements. Different aspects of the management system can be developed by different teams at the same time. The teams can use the same timeboxes and demonstrate working software at regular intervals. As the stakeholders see how the system is evolving, they can make modifications, change the priority, and add new features. There is the option to release increments of the software while continuing to add functionality.
Hired and trained staff	Iterative	Hiring and training staff can use an iterative approach because each of the main functions (winery, hotel, restaurant, etc.) will have their own requirements and their own timing for hiring and training. However, the deliverable won't be considered complete until all functions are staffed and trained. There may be opportunities to learn and adapt hiring and training processes along the way.
Wine club	Incremental	The wine club can use an incremental approach by starting with a minimal set of benefits for members and adding and adapting benefits based on feedback from members.
Grand opening event	Waterfall	The grand opening will need to be planned up front because some of the activities, such as entertainment, will need to be booked in advance. The requirements can be defined at the start, and while there may be minor adjustments, there shouldn't be significant changes. Additionally, the grand opening event has a hard end date; it is not expected to evolve.

Life Cycle

Due to the hybrid nature of the project and the different development approaches, the project will have a start-up and then three main branches each with applicable phases for the type of the work that will be done. Tony has developed a table (Table 5-4) for Tessa that identifies the phase name, work, and completion criteria for each phase.

TABLE 5-4 Life Cycle Phases for Dionysus Winery

Phase	Work	Completion Criteria
Start-up	• Conduct market research. • Concept of operations. • Charter. • Form the team. • Kickoff.	• Market research is favorable for the proposed concept. • A charter is signed by the sponsor and PM. • The team members are identified. • A kickoff meeting is held.
Contracting	• Project procurement strategy. • Request for proposal. • Select source. • Execute contract.	• A comprehensive procurement strategy is developed including contract types. • Statements of work and bid documents are released for each procurement. • Responses are received and proposals are evaluated for each procurement. • Contracts are signed.
Architecture/engineering	• Architectural drawings and elevations; • Engineering specifications.	• All new construction has architectural drawings and elevations. • All renovated buildings have architectural drawings and elevations. • All new construction has engineering plans and specifications. • All renovated buildings have engineering plans and specifications.
Construction	• Foundation. • Framing. • Trade work. • Finish work.	• All construction is permitted. • All construction is consistent with architecture and engineering plans. • Punch list is complete and signed off. • All occupancy permits are issued.

TABLE 5-4 Life Cycle Phases for Dionysus Winery (*Continued*)

Phase	Work	Completion Criteria
Renovation	• Upgrades to current code. • Production equipment ordered and installed. • Storage equipment ordered and installed.	• All upgrades are permitted. • New production equipment ordered, tested, and installed. • Wine storage equipment is ordered, tested, and installed. • Punch list is complete and signed off. • Occupancy permits are issued.
System development	• Identify requirements. • Build, test and demonstrate features. • Integrate and release work.	• Initial development will be considered complete when vineyard data, wine data, inventory management, and wine club features are functional and meet requirements. • After initial release additional features will be the responsibility of operations.
Staffing and training	• Staff positions identified. • Job posting. • Interviews. • Hiring. • Training.	• All staff positions are identified and filled in a timely manner. • All new staff are trained and able to perform their job duties. • After initial staffing, hiring and training will transition to operations.
Wine club	• Market research. • Business model. • Marketing.	• A business model with multiple tiers is approved by Tessa. • Social marketing is launched. • Brochures and flyers are developed. • Managing the wine club is the responsibility of operations.
Grand Opening	• Book entertainment. • Marketing. • Menu/catering plan. • Equipment rental. • Event management.	• A plan for the grand opening is approved by Tessa. • The executive chef has planned and acquired food items for event. • Contracts for equipment, furniture, and materials are in place. • The event is well attended and managed.

After reviewing the table, Tessa asks to see a graphic depiction of how the phases will flow and interact. Tony shows her the information in Figure 5-3.

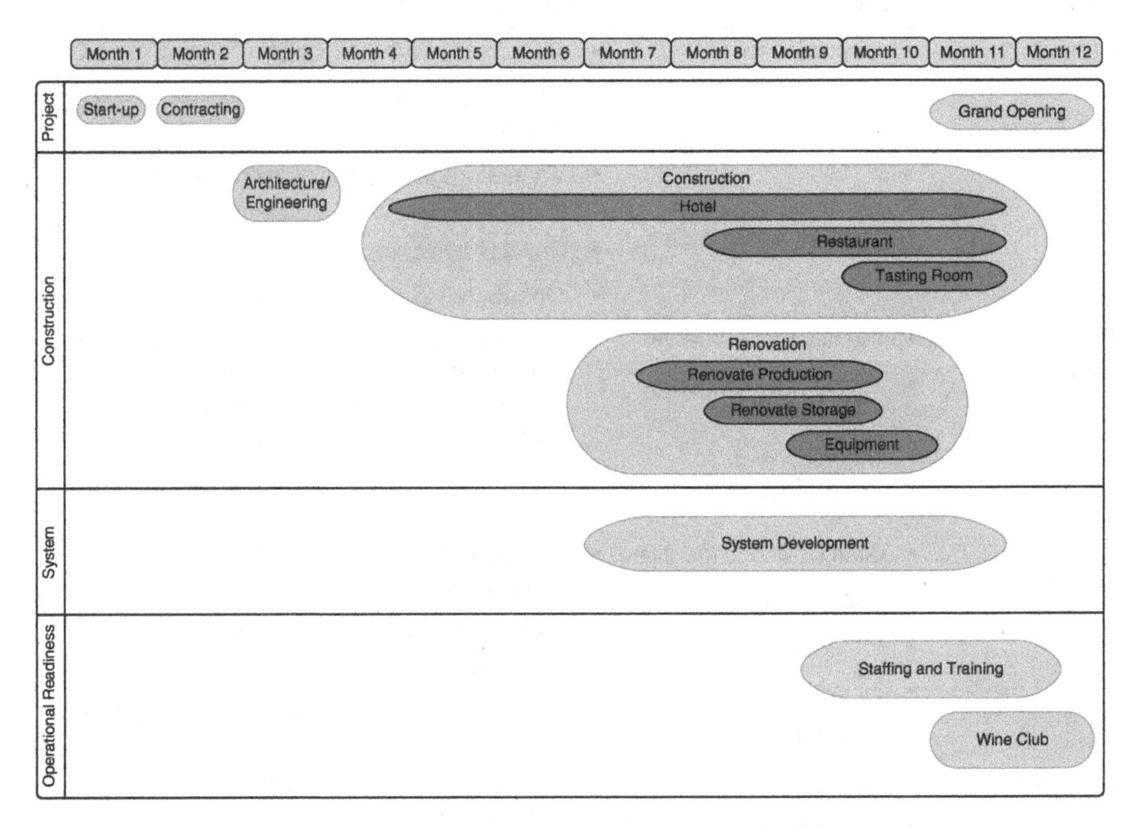

FIGURE 5-3 Dionysus Winery life cycle phases.

Subsidiary Plans

Tony and his team spent several hours discussing how best to manage the project. They evaluated the necessity for control for some deliverables, balanced with the need to be flexible with others. Ultimately, they determined that the project could best be served with the following subsidiary plans:

- Scope management plan;
- Schedule management plan;
- Cost management plan;
- Staffing plan;
- Training plan;
- Construction management plan;
- Safety plan;

- Logistics management plan;
- Communications management plan;
- Risk management plan;
- Procurement management plan;
- Stakeholder engagement plan;
- Change management plan;
- System testing plan;
- Grand opening plan.

Key Reviews

Tessa and Tony met to determine at which points in time Tessa and other members of the management team should review project progress and sign off for advancement. They decided that four reviews would be sufficient.

Integrated baseline review (IBR): The IBR will be used to ensure all the plans are complete, integrated, and incorporate sufficient risk responses. This will take place once the general contractor and key subcontractors are onboard. The review will include construction scope, schedule, and cost baselines along with an integrated risk register. When the IBR is complete, the project will advance into the architecture and engineering phase.

Construction readiness review (CRR): The CRR will give senior management an opportunity to review and approve all blueprints, along with any updates to schedule and cost estimates. Once the CRR is complete, the construction of the hotel, restaurant, and tasting room can begin, along with the renovations to the wine production and wine storage facilities.

System readiness review (SRR): The SRR will be used before taking the inventory, vineyard, wine club, and winery management system live. Because an Agile development approach will be used, stakeholders will see the features as they are developed. The SRR will allow stakeholders to see the entire system integrated and operating in a test environment.

Operational readiness review (ORR): The ORR will include a walkthrough of each venue, a dinner at the restaurant with wine (of course), and a presentation by the managers of each function (hotel, restaurant, winery, facilities, HR, etc.). This review will take place two weeks before the grand opening. This will assure management that all the systems are in place to welcome guests and transition the project to operations.

ROADMAP

> **Project roadmap:** A graphic high-level view of the project that includes phases, reviews, milestones, and other key information.

The project management plan defines the life cycle that shows the general flow of work. It also identifies key reviews. The charter often has key milestones and lists deliverables. However, to

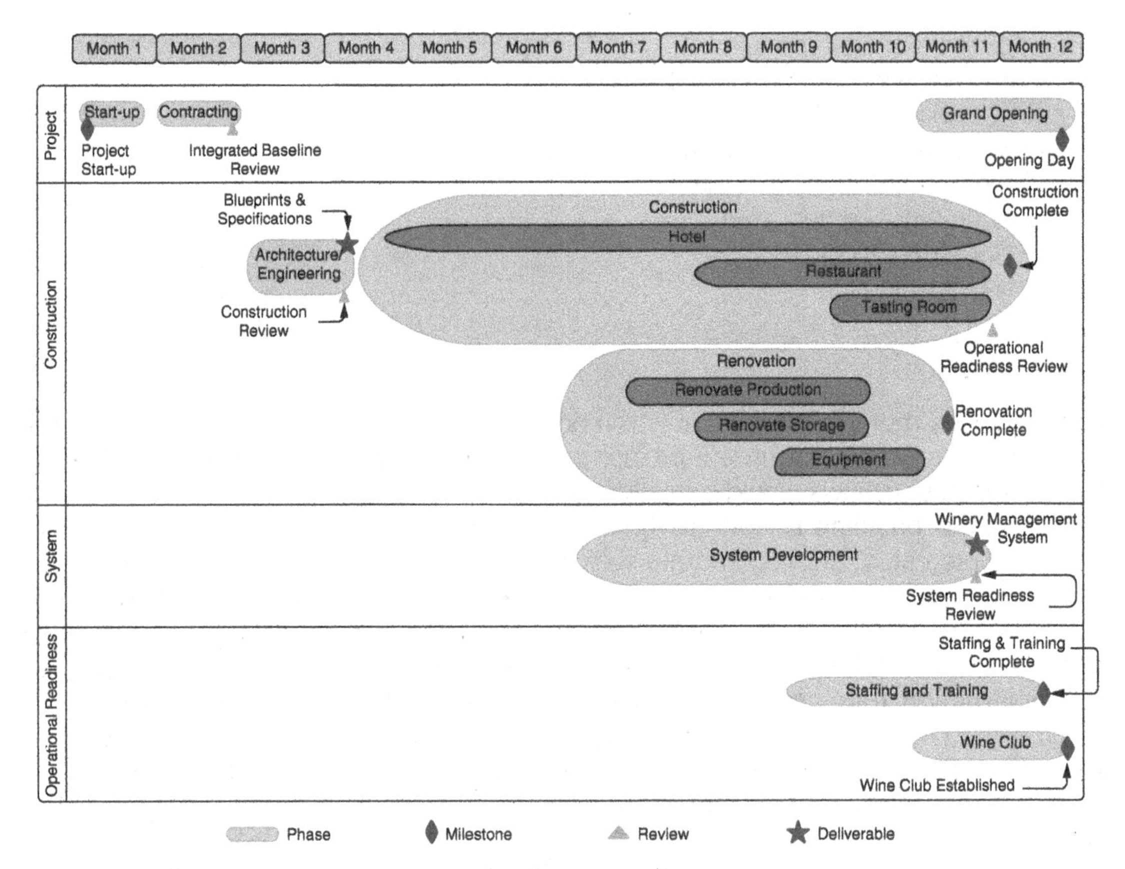

FIGURE 5-4 Dionysus Winery roadmap.

get a high-level view of all the summary information, you turn to a roadmap. In addition to the phases, reviews, milestones, and other key information, it often includes key deliverables, phase gates, and a timeline as well.

Figure 5-4 shows a roadmap for the Dionysus Winery project.

You can see how this roadmap allows the team, senior management, and other key stakeholders to get an overview of the project work and see when major events are planned. Roadmaps come with a legend to identify milestones, phase gates, key reviews, and deliverables.

SUMMARY

In this chapter we identified progressive elaboration and rolling wave planning as key tenets of planning. We described competing demands and how to balance the various project constraints.

We identified elements in the project management plan and described the need to tailor or customize subsidiary plans to meet the needs of the project. We described how the project deliverables influence the development approach, which in turn influences the life cycle phases. Then we covered project reviews. The elements of the project management plan were demonstrated with the Dionysus Winery project.

A project roadmap was introduced as a way of allowing key stakeholders to get an overview of the project work and major events. A roadmap for the Dionysus Winery was shown to demonstrate the usefulness of a roadmap.

Key Terms

phase gate
progressive elaboration
project life cycle
project management plan
project roadmap
rolling wave planning
subsidiary plan

6

Defining Scope in Hybrid Projects

There are many ways to plan and define scope in hybrid projects. You can provide very detailed information with a scope management plan, work breakdown structure dictionary, and requirements software, or you can use a lightweight method by working with user stories and a backlog. Given that scope drives all other aspects of the project, it is important to think about how best to identify, refine, plan, and manage scope for your project.

In this chapter we will start by looking at predictive methods to define scope. These methods work well with deliverables that can be defined up front and will have minimal changes. Then we will look at ways to document and prioritize scope for deliverables with evolving scope. Throughout this chapter we will demonstrate the concepts and methods using the Dionysus Winery project.

PLANNING FOR SCOPE WITH A SCOPE MANAGEMENT PLAN

In Chapter 5 we identified a scope management plan as a subsidiary plan of the project management plan. Here we will look at the scope management plan in more detail.

In a hybrid project you will have some deliverables that will use predictive development approaches and some that will evolve the scope as the project progresses. You can use the scope management plan to document how you plan to work with both types of scope definition and refinement. For example, you may want to have a section in your scope management plan for predictive scope and a section for adaptive scope.

Scope management plan: A subsidiary plan of the project management plan that describes how scope will be defined, documented, managed, and verified.

The section on predictive scope may document how you will decompose scope from the charter, organize the scope, control changes to scope, and verify that the scope is correct. The section on adaptive scope may identify how you will prioritize and validate scope.

A scope management plan may also document how scope management and requirements management will integrate. If significant deliverables are being developed by contractors, the plan may discuss means for integrating procured scope and scope that is developed in-house.

Below you can see an example of a scope management plan for the Dionysus Winery project.

Scope Management Plan

Predictive Scope

Decompose

Those elements that are predictive in nature, such as the hotel, restaurant, tasting room, wine production facility, wine storage facility, and the grand opening, will be further decomposed using a scope statement. The scope statement will provide a narrative description of each deliverable, identify components associated with each of the deliverables, and identify out-of-scope work (exclusions).

Organize

A work breakdown structure will be used to organize the scope and further decompose the scope into low-level deliverables. Progressive elaboration will be used to further decompose the work as more information becomes available.

A WBS dictionary will be used only for the grand opening. The general contractor will manage the detailed information for all the construction and renovation work.

Control Changes

A change control board (CCB) will be established to ensure that scope remains tightly controlled. Members of the change control board include Anthony Dakota, Tessa Barry, and the general contractor. If needed additional people will be brought in to help the CCB understand the nature and implications of proposed changes.

Any suggested changes will require a change request, justification, and estimate of the impact on the competing demands of scope, quality, schedule, cost, resources, and risk.

A change log will be kept that documents the information from all change requests and how they were addressed.

The change control board will meet every two weeks to review change requests. Their decision will be documented in the change log and then communicated to relevant stakeholders.

Verification and Validation

Verification (technical correctness) and validation (customer acceptance) will be conducted throughout the project. For construction work, verification and validation will consist of:

Sign-off on requirements documentation
Sign-off on blueprints
Walk through when all trade work is complete
Walk through when all finish work is complete
Review of all building permits
Review of occupancy permits
Successful reviews (integrated baseline review, construction review, and operational readiness review)

Adaptive Scope

The adaptive scope is composed of hiring and training, establishing the wine club, and developing the winery management system.

Decompose

The hiring and training along with the wine club will be decomposed using a separate scope statement for each deliverable. The scope statement will provide a narrative description and a description of associated deliverables and will identify out-of-scope work (exclusions).

Organize

A resource breakdown structure will be used to organize the staffing, and a training plan will be used to organize and structure the training.

The winery management system and wine club will be organized and prioritized using a separate backlog for each.

Control Changes

Based on adaptive principles, changes can be introduced; however, the due dates are inflexible. Schedule delivery is paramount; therefore, any modifications to scope must not interfere with timely completion.

Verification and Validation

Hiring and training will be verified and validated by corporate HR based on an audit of the hiring and training process.

The wine club will be validated by Tessa Barry.

The winery management system will be verified and validated with regular demonstrations of functionality. These demonstrations will take place at a regular cadence and will be attended by Anthony Dakota, Tessa Barry, the operations manager (to be hired), and the corporate IT director. Final acceptance is the responsibility of the operations manager.

Smaller projects may not always use a scope management plan. However, because hybrid projects are by their nature more complex, the plan can be useful to differentiate how scope will be managed for the various aspects of the project.

ELABORATING SCOPE WITH A SCOPE STATEMENT

A scope statement uses the information from the project charter and progressively elaborates it. The scope statements contains a narrative description of the project, a description of each deliverable, and identifies work that is out of scope. If acceptance criteria aren't identified elsewhere (such as the scope management plan or a verification and validation plan), it is identified in the scope statement.

Scope statement: A document that describes project scope and deliverables and identifies out-of-scope work.

Narrative Description

The narrative description paints a picture of the end result in more detail than the charter. It gives a richer representation of the project overall. For example, the Dionysus Winery vision statement is:

For adults who appreciate wine and good times, the Dionysus Winery is a state-of-the-art winery that caters to the fun and frivolity in all of us. Unlike other wineries that only offer wines, clubs, and tasting rooms, the Dionysus Winery will offer education and recreation activities, unique lodging, and innovative, locally sourced gourmet food.

The purpose, as documented in the charter, is:

Provide a fun, upscale, unique winery experience for individuals, couples, and groups.

The narrative description in the scope statement could be:

> The Dionysus Winery will be a boutique winery that will provide a fun and engaging experience for patrons. Guests will have the opportunity to learn about grapes, the wine making process, taste wines in various stages of fermentation and aging, and taste how different foods and wines complement one another.
>
> In addition to learning about wine, Dionysus Winery will offer fun events such as balloon rides, music, grape stomps, horseback rides in the vineyards, and other similar events.
>
> Our restaurant and tasting room will provide locally sourced food and have the ambiance of a Tuscan winery. The grounds will provide romantic venues for weddings, anniversaries, and other special events.

You can see how this narrative provides a richer description and deeper understanding of the project.

Deliverables

Each deliverable is elaborated to communicate a more complete concept of the deliverable. The process of elaborating each deliverable enables the team to start making decisions, reduce uncertainty, and shape the deliverables. Often the team needs to complete further analysis and research as part of completing the scope statement. This may include creating a decision matrix or breaking the deliverables down via a system engineering perspective, value analysis, product breakdown structure, or other means of identifying the component parts of the project.

An example for the Dionysus Winery project demonstrates the process of elaborating the grand opening event. The team was considering three options for the grand opening: a "Taste of Tuscany" buffet, a sit-down dinner with complementary wine, or a grape stomp. Tessa prioritized her desired outcomes for the event as:

1. The more people the event can accommodate, the better;
2. The event should be memorable—something people don't normally do;
3. The cost should be reasonable.

The team used a decision matrix to determine the best option. A decision matrix evaluates multiple options by creating a matrix with a set of criteria important to the decision. The matrix lists the options in the left column and the criteria across the top row. The criteria may be weighted to indicate priorities. The options are scored against each criterion, and the weights are multiplied by the scores to determine the optimal outcome.

Decision matrix: A tool used to evaluate multiple options against a set of criteria.

Using the criteria Tessa identified and the three options, the team created a decision matrix. They weighted the importance of the criteria based on Tessa's preference: the option that best fulfilled the criterion would be rated as a 5, the one that next best fulfilled it would be rated as a 3, and the one that least fulfilled the criteria would be rated as a 1. Table 6-1 shows the decision analysis.

Based on this analysis, it was obvious that the theme of the grand opening event should be a grape stomp.

TABLE 6-1 Grand Opening Decision Analysis

		Stomp		Sit-down dinner		Taste of Tuscany	
Criteria/options	**Weight**	**Rating**	**Score**	**Rating**	**Score**	**Rating**	**Score**
Lots of people	40%	5	2.0	1	.4	3	1.2
Memorable	35%	5	1.75	3	1.05	1	.35
Reasonable cost	25%	3	.75	1	.25	5	1.25
Total			4.5		1.7		2.8

A deliverable description for the grand opening could be documented as:

> The grand opening event will provide a fun and memorable grape stomping event. People will compete in teams of two with one person stomping and another catching the juice in a bottle. Each "heat" will have 12 teams. When all the teams have competed, the winner from each of the heats will have a stomp-off. The grand winner will get a free wine club membership for a year.
>
> The grand opening will showcase all Dionysus wines at multiple tasting stations and have big wheels of cheese and mounds of crackers to go with the wines. A band will play before and after the stomp event.

Out of Scope

When elaborating and documenting scope for a project, it is just as important to identify those areas that are out of scope as it is to identify those areas that are in scope. By specifically identifying out-of-scope work up front, you can manage stakeholder expectations and address any concerns about in and out of scope before creating a baseline.

At the planning meeting the team identified the following areas as out of scope for the Dionysus Winery project.

- Any work and costs associated with fixing up and managing the vineyard;
- All maintenance and operations after the grand opening;
- Negotiating salaries;
- Staff labor costs are not included in the budget.

ORGANIZING SCOPE WITH A WORK BREAKDOWN STRUCTURE

The scope statement can help the team elaborate the project scope and each deliverable. However, you need a way to help organize the scope. A work breakdown structure (WBS) can help organize and define the total scope of the project—both product work to create deliverables and project work to organize and manage the project.

Work breakdown structure (WBS): A tool used to decompose and organize project and product scope.

When you start the WBS, it looks like an org chart. It starts at a high level and progresses downward as deliverables are broken down into finer levels of detail. In addition to organizing the work, the WBS also enables the team to identify all the deliverables necessary to meet the project objectives. In other words, the WBS contains all the work in the project. If the work isn't represented in the WBS, it is out of scope. Figure 6-1 shows a high-level WBS for the Dionysus Winery project.

Tip: When working with a WBS, the terms "breaking down," "decomposing," or "deconstructing deliverables" all mean the same thing.

WBS Levels

There are some nomenclature concepts you need to know about a WBS. Each level has a number. The project is level 1, in this case Dionysus Winery. The top-level organization

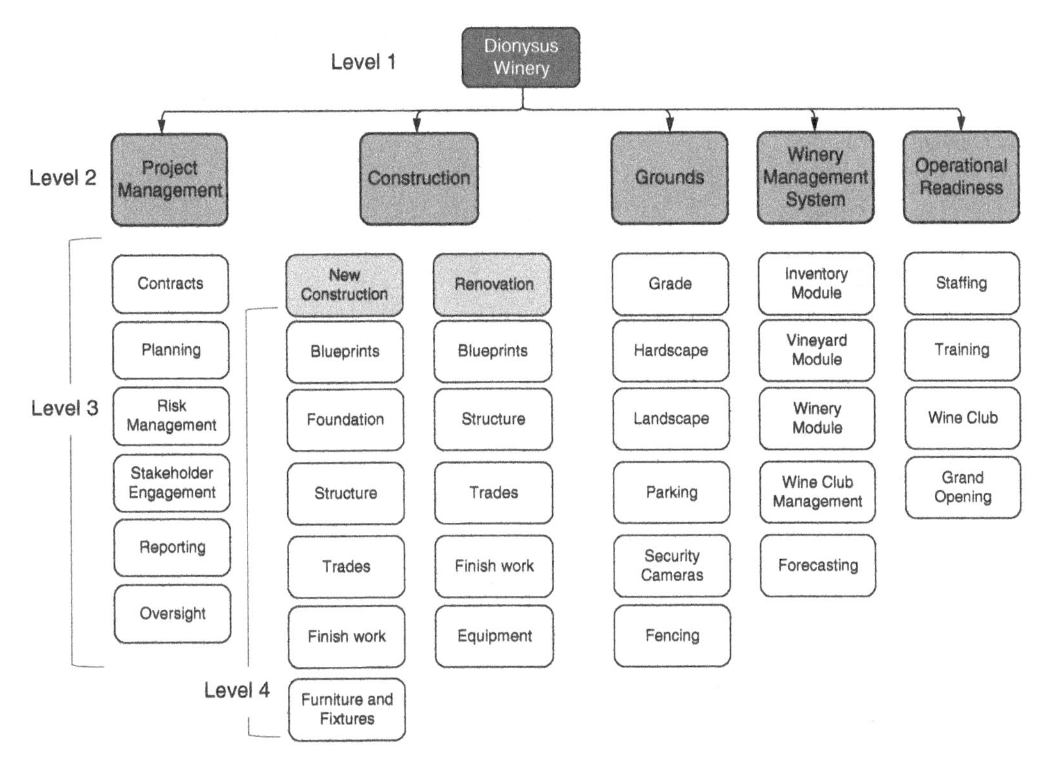

FIGURE 6-1 Dionysus Winery WBS.

is level 2 (Project Management, Construction, Grounds, Winery Management System, and Operational Readiness). Each lower level of decomposition follows from there, level 3, level 4, and so forth.

The WBS in Figure 6-1 follows a best practice of organizing the work by deliverable. Notice that other than project management, WBS elements are deliverables (nouns). Grouping by deliverable also means that the WBS is arranged by logic, not sequence. The work will be further decomposed into activities and put in sequence when it is scheduled.

Some projects use level 2 to organize by life cycle phase rather than deliverable. Projects that have multiple locations may arrange the work by geography. While arranging the WBS by major deliverables is a best practice, it is not the only way—as with everything in project management—so tailor your approach to meet the needs of your project.

The WBS will be progressively elaborated just like most project artifacts. As the project progresses, it makes sense to provide more detail. At some point working with a chart like the one in Figure 6-1 gets unwieldy. Usually when the project gets beyond level

3 or level 4 it is documented in outline format. The portion of the WBS from Figure 6-1 is shown below in an outline format.

1. Dionysus Winery
   1.1. Project Management
      1.2.2. Contracts
      1.2.3. Planning
      1.2.4. Risk Management
      1.2.4. Stakeholder Engagement
      1.2.5. Reporting
      1.2.6. Oversight
   1.2. Construction
      1.2.1. New Construction
         1.2.1.1. Blueprints
         1.2.1.2. Grade
         1.2.1.3. Foundation
         1.2.1.4. Frame
         1.2.1.5. Roof
         1.2.1.6. Trades
         1.2.1.7. Finish Work
         1.2.1.8. Furniture and Fixtures
      1.2.2. Renovation
         1.2.2.1. Blueprints
         1.2.2.2. . . .

Since the construction scope is very predictable, that work can be decomposed to a low level at the start of the project. However, the winery management system is evolutionary; therefore, it only makes sense to decompose it to level 3, with the key functions we know about: the inventory module, the vineyard module, and the winery module. The wine club is using an incremental approach; therefore, that branch of the work will be decomposed as the team gets feedback on what the stakeholders want.

Work Packages, Planning Packages, and Control Accounts

For smaller projects that only require three to four levels, the information presented about the WBS thus far is sufficient. However, for a large project like the winery, you will also want to have some terminology to describe the level of detail. There are three levels of detail in a WBS, a work package, a planning package, and a control account.

A work package is the lowest-level deliverable in a WBS. It cannot be decomposed any further without listing the activities (verbs) associated with the work. A WBS is not used to represent the activities, only the deliverables. The WBS in Figure 6-1 is not decomposed to show the work packages. If we were to decompose the roof into work packages, it might look like this:

> **Work package:** The lowest-level deliverable in a WBS.

1.2.1.5. Roof
 1.2.1.5.1. Roof frame
 1.2.1.5.2. Insulation
 1.2.1.5.3. Flashing
 1.2.1.5.4. Shingles
 1.3.1.5.5. Drains
 1.2.1.5.6. Scuppers
 . . . .

A planning package is for work that hasn't been decomposed into work packages. This can be due to not having enough information to decompose the work into work packages, employing rolling wave planning, or an approved change request that hasn't yet been decomposed. The WBS in Figure 6-1 shows mostly planning packages that have not yet been fully decomposed.

> **Planning package:** A component in the WBS that represents work that hasn't been decomposed into work packages.

A control account is used for control and reporting purposes. A control account is held at a high level on the WBS. It encompasses the planning packages and work packages beneath it. Each control account has multiple work packages, and each work package only belongs to one control account. A control account is used to measure cost and schedule performance. When reporting to senior management, they usually don't want to hear about every single work package, especially if there are no performance issues. Therefore, you "roll up" the performance information in a way that makes sense for the project.

> **Control account:** A component of the WBS used to control work performance and report on cost and schedule status.

For the winery project it might make sense to report performance for the Grounds, for the Operational Readiness, and the Renovation. Each of these would be control accounts, and the performance for all the work underneath them would be combined for reporting purposes.

For the new construction, Tessa and Anthony will have to determine what level of detail she wants. If Tessa wants high-level information, then reporting on the new construction is fine. If she wants more detail, then it might make sense to restructure that branch of the WBS so it is organized the way progress will be reported.

Steps to Create a WBS

To create a WBS, follow these five steps:

1. Analyze the scope statement and requirements to understand the project work;
2. Determine how you will structure the WBS (by major deliverable, phase, etc.);
3. Decompose the work until you have work packages or until you can't realistically decompose it further;
4. Identify control accounts (if using control accounts for reporting purposes);
5. Establish a numeric coding structure, especially if your WBS has more than 50 work packages.

GETTING INTO THE DETAIL WITH A WBS DICTIONARY

WBS dictionary: A document that provides a definition of work, activities, milestones, resources, costs, and other information for components of a WBS.

The WBS dictionary is used on larger projects to provide greater detail for components in the WBS. The WBS dictionary includes a definition of the scope or statement of work, defined deliverable(s), a list of associated activities, and a list of milestones. Other information may include start and end dates, resources required, an estimate of cost, contractual information, quality requirements, and technical references to facilitate performance of the work. Figure 6-2 shows an example of a WBS Dictionary template.

If you don't need as much detail as the example shown in Figure 6-2, you can tailor the form to meet your needs. You won't need a WBS dictionary for all projects, but a WBS dictionary is helpful as a communication tool for large projects to make sure that all the work details associated with a deliverable are fully understood.

WORKING WITH REQUIREMENTS

Requirement: A capability that must be present or a condition that must be met to achieve the project objectives.

Part of understanding and documenting scope is working with requirements. All hybrid projects have requirements, even if they are called stakeholder needs, performance criteria, or some other name. Requirements management is a discipline unto itself, so we won't go into significant detail here, but we will discuss requirements in general.

WBS DICTIONARY

Project Title: ______________________ Date Prepared: ______________________

Work Package Name:	Code of Accounts:
Description of Work:	Assumptions and Constraints:
Milestones: 1. 2. 3.	Due Dates:

ID	Activity	Resource	Labor			Material			Total Cost
			Hours	Rate	Total	Units	Cost	Total	

Quality Requirements:

Acceptance Criteria:

Technical Information:

Agreement Information:

Page 1 of 1

FIGURE 6-2 WBS dictionary template.

Elicitation

The first aspect of requirements management is eliciting requirements. Eliciting requirements includes more than just asking stakeholders what their requirements are. There are several techniques you can use to elicit requirements. I have grouped them into three main categories: working with stakeholders, research, and visual methods.

> **Elicitation:** A structured approach to draw out requirements.

Working with Stakeholders

A good first step to identify and come to an agreement about requirements is to work with project stakeholders. This can include end users, the customer, support staff, team members, vendors, and anyone else who can help you identify requirements. Some of the common methods of working with stakeholders are described below.

Interviews: Interviews include asking people how they will use the end product, problems they currently have that the product will solve, and features and functions they want to see and those they don't want to see. One of the powerful aspects of interviews is that you can ask open-ended questions. By listening, probing, and asking for clarification you can learn a lot about what people want and need.

Focus groups: Focus groups are similar to interviews, but they are conducted by a trained facilitator. The facilitator works with a group of stakeholders. The facilitator may present scenarios, ask open- or close-ended questions, and leverage group discussions to find out about obvious and not so obvious needs.

Brainstorming: Brainstorming can be very productive when performed with stakeholders from different functions. For example, at the Dionysus Winery the operations staff, financial manager, vineyard manager, and wine production staff will all have different needs. Rather than talking with them individually, it can be very productive to have them all in the same room brainstorming what they need.

Surveys and questionnaires: Surveys and questionnaires are a good way to collect data from many different stakeholders quickly. Many questionnaires provide multiple-choice questions that can be quickly tabulated to quantify results.

Prototypes: Prototypes provide a functional or nonfunctional model of a product or service. They allow stakeholders to see how a system or product will operate. Prototypes can be as simple as a visual model that is used to gather feedback and recommendations. For example, you might have a model of the winery grounds that shows where the vineyards, production facility, storage, restaurant, hotel, and parking are located. This can help people visualize the end result, ask questions, and make suggestions.

Research

Research methods can include reviewing academic papers, market research, benchmarking, document analysis, and observation. Academic papers and market research are self-explanatory. Benchmarking, document analysis, and observation are described below.

Benchmarking: Benchmarking includes identifying the best in class or best in industry examples and using those as a target to achieve or surpass. It can include identifying effective processes and best practices.

Document analysis: Reviewing documents can uncover requirements that stakeholders may not think of. Document analysis can include reviewing error logs, complaint files, warranty work, schedules, budgets, and so forth.

Observation: Sometimes the best way to understand requirements is to see how people interact with a similar product. You can observe workflows, user interactions, and the environment. This allows you to see how people engage with a system or deliverable, rather than how they say they engage with it. For the winery, you can observe how visitors to other wineries behave, where they spend their time, what they are excited about, and what they avoid.

Visual Methods

The two main visual methods are mind mapping and affinity diagrams.

Mind mapping: A mind map is a visual representation of brainstorming. The central idea is written in the center, and then different categories of requirements branch out from the center (see Figure 6-3). Each branch may have offshoots with requirements or ideas. The mind map organizes the requirements and thoughts as they are identified. You can use a paper mind map with sticky notes, or an electronic mind map to capture ideas. You can see a mind map for the Winery Management System in Figure 6-3.

Affinity diagrams: An affinity diagram uses a technique called affinity grouping to categorize ideas based on similarities (see Figure 6-4). It is similar to a mind map but is set up as a table or breakdown structure. Figure 6-4 shows an affinity diagram for the Winery Management System.

> **Affinity grouping:** Categorizing elements into groups based on similar characteristics.

Prioritization

Once requirements are identified, they should be prioritized. One way to prioritize is by function, such as "inventory requirements are most important, followed by production, security, and so forth." Another option is by schedule. For example, the vineyard operations will be active first, followed by production, storage, inventory, and then administration.

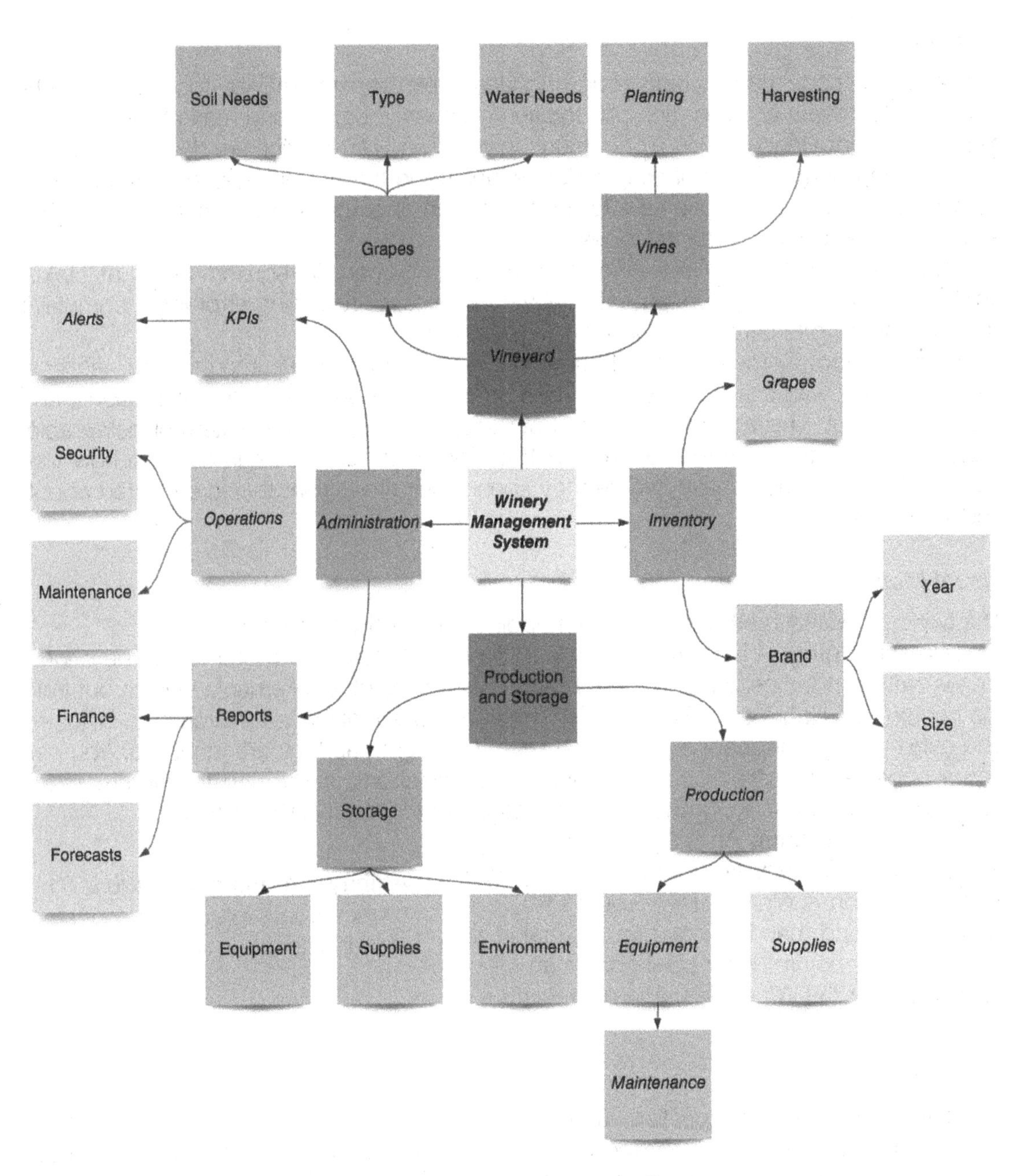

FIGURE 6-3 Mind map for the Winery Management System.

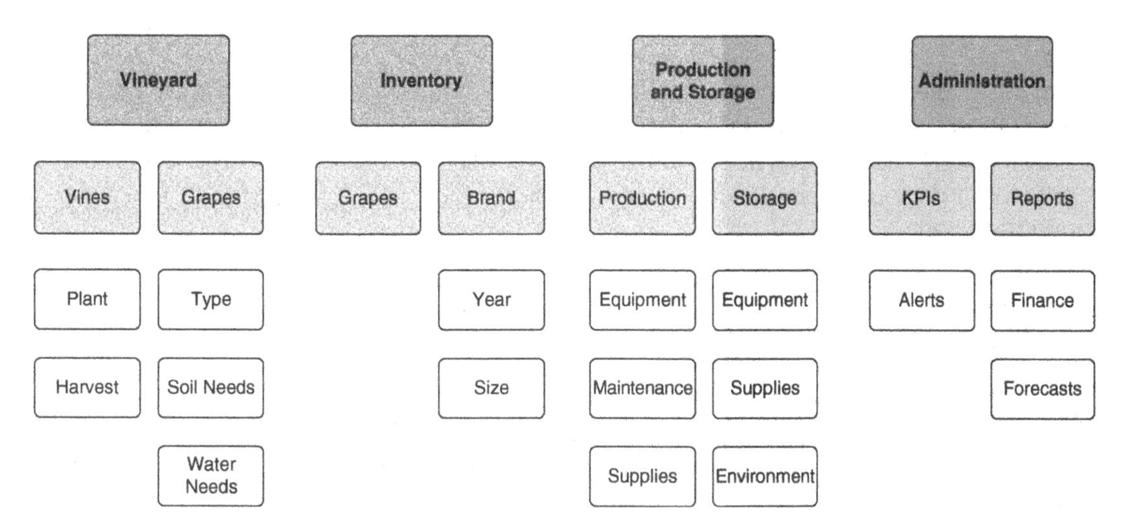

FIGURE 6-4 Affinity diagram for the Winery Management System.

You can also have stakeholders participate in the prioritization by using a nominal group technique. To use the nominal group technique for requirements prioritization, follow these steps:

1. Brainstorm requirements;
2. Discuss the requirements so everyone has a common understanding;
3. Have every person rate each requirement. Here are three different examples you can use for rating:
   a. Each person numbers the requirements from most important to least important;
   b. Each person has a set number of points (such as 100) that they allocate based on their opinion of the relative importance of each requirement;
   c. Each person assigns a number from 1 to 5 based on their priority: 1 is not important, and 5 is critically important.
4. Tally the results.

A simplified version of nominal group technique is dot voting. With dot voting, each person has a set number of dots. They allocate their dots to the requirements they think are most important. When all the dots are placed, they are tallied.

Documenting Requirements

There are several ways you can document requirements. Large projects with hundreds or even thousands of requirements use software to record and manage requirements. However, you can also use low-tech means, such as index cards or user stories.

Cards

A card is a reference to when requirements were recorded on index cards and catalogued. Today we generally use electronic cards due to the increasing remote work and distributed teams. Cards collect basic information about the requirements and record the information in a consistent fashion. Figure 6-5 shows an example of requirement card.

Requirement #:	Category:
Description:	
Originator:	Priority
Acceptance criteria:	
Comments:	

FIGURE 6-5 Requirement card.

User Stories

A user story describes what a stakeholder wants something to do and the benefit it provides. They are commonly phrased as:

"As a ____________, I want ____________ so I can ____________."

User story: A brief description of a desired outcome, documented from a stakeholder's perspective.

This method keeps the stakeholder's needs visible throughout the development process. For the winery you might see the following user stories:

> As the winery production manager, I want to track equipment maintenance needs so I can schedule and maintain the equipment efficiently.
>
> As the operations manager, I want to track the wine inventory by brand so I can ensure we have sufficient inventory to meet demand.
>
> As a wine club member, I want to know the history of each bottle of wine so I can increase my knowledge of wine.

PRIORITIZING SCOPE WITH A BACKLOG

The WBS and WBS dictionary are used with predictive projects. For projects using an Agile methodology with evolving scope, a prioritized backlog is used. Figure 6-6 shows part of a backlog for the winery management system.

Backlog: A list of work to be done.

The backlog is owned and prioritized by the product owner. The scrum master assists the product owner with keeping the backlog up to date and clarifying the work with the team. The product owner may add new work or reprioritize existing work based on a stakeholder request, stakeholder feedback, updated information,

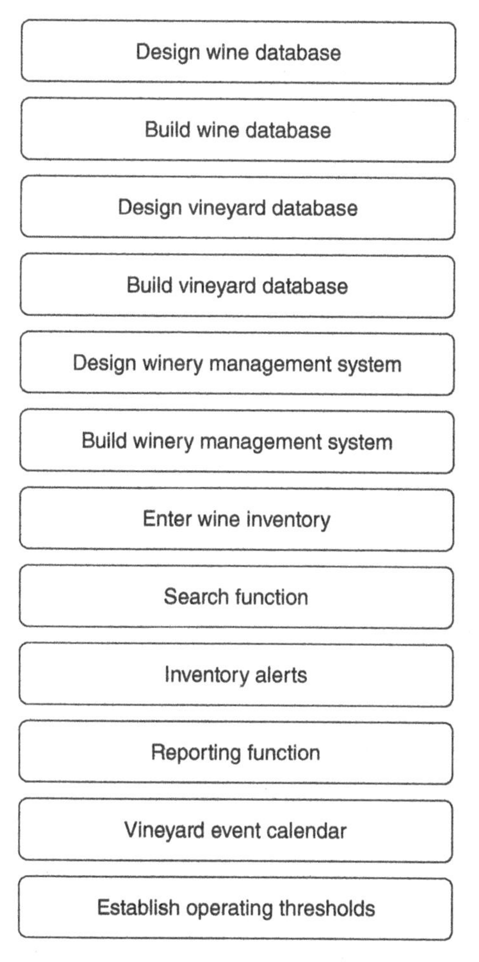

FIGURE 6-6 Winery management system backlog.

and so forth. A backlog can be documented on a flipchart, a shared platform, or with sticky notes.

Note: You can create a backlog with features, requirements, or user stories.

SUMMARY

In this chapter we described how a scope management plan can define how scope will be defined, documented, and managed. For projects with predictive scope, we discussed using a scope statement to elaborate information from the project charter, and a WBS to organize the scope. For large projects, we described using a WBS dictionary as a structure to record deliverable details.

Scope and requirements go hand in hand. We described ways to elicit, prioritize, and document requirements. We concluded by looking at how a backlog can be used to evolve adaptive scope and prioritize stakeholder needs.

Key Terms

affinity grouping
backlog
control account
decision matrix
elicitation
planning package
requirement
scope management plan
scope statement
user story
WBS dictionary
work breakdown structure (WBS)
work package

7

Building a Predictive Schedule

One of the most common questions we are asked as project managers is, "When will it be done?" Sometimes it seems like that question is the adult version of your kids asking, "Are we there yet?" In all seriousness, your project schedule is the most comprehensive view of project progress. It contains all kinds of information from milestones to resources to start and end dates and more. Knowing how to develop, optimize, and maintain your schedule is a fundamental requirement for project success.

In this chapter we'll discuss how you can use a schedule management plan to help plan how you manage the predictive and adaptive aspects of your schedule. Then we will describe four steps for building a schedule with well-defined scope.

ORGANIZING WITH A SCHEDULE MANAGEMENT PLAN

Schedule management plan: A subsidiary plan of the project management plan that describes how the schedule will be developed, managed, and maintained.

A project where some deliverables can be well defined with reasonably accurate duration estimates can be scheduled with a fair degree of confidence and accuracy. The same can't be said for projects where some of the deliverables can't be defined or agreed to until part way through the project. The schedule approaches for these two situations are very different. But on a hybrid project you still have to put together a schedule that accounts for both situations. This is where a schedule management plan can help you think through the best way to develop and maintain a project schedule.

Like all project artifacts, a schedule management plan should be tailored to meet your needs. Table 7-1 describes information commonly found in a schedule management plan.

TABLE 7-1 Schedule Management Plan Contents

Schedule methodology	Document the scheduling methodology that will be used for the project, including critical path, Agile, or other methodology.
Scheduling tool	Identify the scheduling tool(s) that will be used for the project: scheduling software, task boards, and so forth.
Level of accuracy	Describe the level of accuracy needed for estimates. If there are guidelines for rolling wave planning and the level of refinement that will be used for duration and effort estimates, then you will indicate the levels of accuracy desired as time progresses.
Units of measure	Indicate whether duration estimates will be in days, weeks, months, story points, or some other unit of measure.
Variance thresholds	Indicate acceptable variances for work to be considered on time, in jeopardy, or late.
Schedule updates	Document the process for updating the schedule, including update frequency and version control. Indicate the guidelines for maintaining baseline integrity and for re-baselining if necessary.

For the Dionysus Winery project, the schedule management plan will include information about how the deliverables, which use multiple development approaches and different scheduling methods, will be incorporated into one schedule. There are several scheduling terms in the Dionysus Winery schedule management plan that have not been defined or used thus far. Those terms are defined below.

Integrated master schedule: A schedule that combines all schedules for a project into one all-encompassing document.

Critical path: The series of tasks through a project with the longest duration, which determines the soonest the project can be completed.

Critical path methodology: A scheduling approach that identifies the path that drives project duration and identifies the amount of scheduling flexibility on other paths.

Summary task: A task that aggregates information from detailed work into a single task.

Release. A set of features, functions, or deliverables that are placed into use.

Release plan: A plan that shows the expected timing, milestones, and outcomes for releases.

Iteration: A brief, set time interval in a project where the team performs work. Also known as a timebox.

Story points: A relative unit of measure used to estimate work in a user story.

The schedule management plan for the Dionysus Winery project is shown below.

Schedule Management Plan

Schedule Methodologies and Tools

Because there are different development approaches that will be used in this project, we will use an integrated master schedule (IMS) that can accommodate both critical path and iterations.

Those elements that are predictive in nature—the hotel, restaurant, tasting room, wine production facility, wine storage facility, and the grand opening—will use critical path methodology (CPM).

Hiring and training will initially be documented as a summary task. It will use an iterative approach where each function will use the same set of activities. The duration and start and end dates will be defined and refined by the responsible functional manager.

The initial set of benefits for the wine club will be established as part of the project. Because the benefits are not yet defined, the wine club will be documented as a summary task that will be elaborated as information becomes available.

The winery management system will be developed using Agile methodology. A release plan will be developed using a series of two-week iterations. The first release will be a minimum viable product. Thereafter the product owner (operations manager) will prioritize additional features and functions. The integrated master schedule will show the 2-week iterations. The detail will be managed using a task board.

Level of Accuracy and Units of Measure

Work that uses CPM will be planned out in full. Work occurring within a 60-day window will be refined to show duration in days. Work occurring from 60 days to the end of the project will be estimated in weeks.

Iterative and incremental work will use weeks as a unit of measure.

Agile work will use fixed duration timeboxes.

> **Variance Thresholds**
>
> Variances of greater than one week on the critical path or near-critical paths are considered late. Variances on noncritical paths that use more than 50% of their float are considered in jeopardy.
>
> **Schedule Updates**
>
> The integrated master schedule will be updated with progress on a weekly basis. Schedule status will be reported to the sponsor monthly.

Many projects don't need a schedule management plan. But because hybrid projects have different methods of scheduling and may use different schedule tools, it is a good idea to think through how to integrate that information. Having an integrated master schedule promotes efficient schedule planning, manages stakeholder expectations, and allows you to determine overall project status.

PREDICTIVE SCHEDULING

Developing a schedule for those parts of the project that have clear and fixed scope involves six steps:

1. Identify tasks;
2. Sequence tasks;
3. Assign team members;
4. Estimate durations;
5. Analyze the schedule; and
6. Finalize the schedule.

The next several sections cover this steps 1 through 4 in more detail. Steps 5 and 6 are in the next chapter.

Identify Tasks

> **Task:** A distinct element of scheduled work. Also known as an activity.

Tasks are the actions needed to create the work packages. When you create your WBS, you decompose the work to create work packages. For creating the schedule, you decompose work packages to create tasks. In other words, tasks are verbs. Your WBS should be composed of nouns, and your task list should be composed of verbs.

Figure 6-1 in Chapter 6 shows a high-level WBS. The task list below elaborates the work package for the foundation. You can see that the numbering scheme developed for the WBS in the list format is continued for the tasks.

1.2.1.3. Foundation
  1.2.1.3.1. Excavate
  1.2.1.3.2. Grade
  1.2.1.3.3. Install footings
  1.2.1.3.4. Place forms
  1.2.1.3.5. Install rebar
  1.2.1.3.6. Pour concrete
  1.2.1.3.7. Remove forms

For short projects you may not need to use rolling wave planning. However, for projects longer than six months, identifying all the tasks that need to take place usually isn't feasible. Therefore, identify only those tasks you need to accomplish in a 60–90-day range, and keep the future work at a higher level.

> **Tip:** The words "activity" and "task" are interchangeable. Most scheduling software uses the term "task." *A Guide to the Project Management Body of Knowledge (PMBOK® Guide)* uses the term "activity."

It is a good practice to ask your team members to identify the tasks needed to complete the work packages. They have the subject matter expertise necessary to identify all the work needed. If you are able to start with a schedule from a similar project, you will still need to ask your team members to help you tailor it to meet the needs of your project.

Sequence Tasks

Once you have identified your tasks, you will need to put them in order. Most work on a project occurs sequentially—one task ends, and then the next one starts. However, there are some situations where tasks start at or around the same time or finish at or around the same time. The three main relationships between tasks are:

- **Finish-to-start:** The preceding task finishes and then the next task starts. The tasks in Figure 7-1 all have a finish-to start relationship. A finish-to-start relationship is abbreviated as FS. This is the most common relationship.

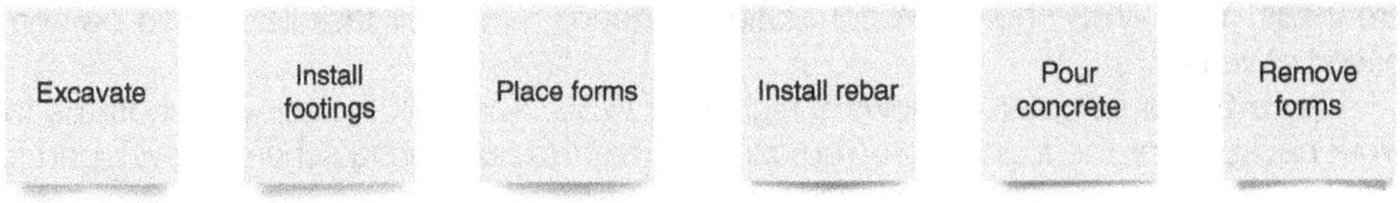

Excavate | Install footings | Place forms | Install rebar | Pour concrete | Remove forms

FIGURE 7-1 Network diagram with sticky notes.

- **Start-to-start:** This relationship occurs when one task starts, and then the next task starts. In this situation the preceding task only needs to start before the next task starts, it doesn't have to finish before the next task starts. For the grand opening event for the winery, the task of planning the menu will start before planning the wine list. The menu doesn't have to be fully planned out before the wine list is created, but it should at least be started. A start-to-start relationship is abbreviated as SS.
- **Finish-to-finish:** The preceding task must finish before the next task can finish. For example, if the operations manager is putting together an operations manual, he can write the first draft of the content and then give it to a professional editor who will edit and format the content. For this series of tasks, some of the work can occur concurrently. However, the editing can't finish until the writing is finished. The formatting can't finish until the editing is finished. This type of finish-to-finish relationship is abbreviated as FF.

Finish-to-start: A relationship where the preceding task finishes before the next task can start.

Start-to-start: A relationship where the preceding task starts before the next task can start.

Finish-to-finish: A relationship where the preceding task finishes before the next task can finish.

Tip: There is a start-to-finish (SF) relationship, but these are rare. An example would be a new system must be running before the previous system can finish shutting down. This type of relationship is best left to professional schedulers.

It is a useful practice to use sticky notes to work out the initial sequence of tasks. That way if you make a mistake, find a better way, or forget a task, it is easy to rearrange the sticky notes.

Leads and Lags

Sometimes you don't want to start a task as soon as its predecessor finishes—sometimes you want to wait. In scheduling terminology this is called a lag. For example, using the tasks for the foundation work package, after you pour the concrete (Task 1.2.1.3.6), you will have a lag of 10 days before you remove the forms (Task 1.2.1.6.7). A lag is indicated with a plus sign (+). Thus, the relationship between these two tasks is shown as FS + 10d, where "d" indicates days.

Lag: A delay between tasks.

Lead: An acceleration between tasks.

If you want to accelerate the time between two tasks, you use a lead. For example, if you have an FS relationship between gathering requirements and designing, you could apply a lead to accelerate the design. This is represented as FS – 2w, where FS is the finish-to-start relationship, the lead is shown as a minus sign, and 2w signifies two weeks.

Table 7-2 shows some of the abbreviations used in scheduling.

TABLE 7-2 Scheduling Abbreviations

Abbreviation	What it means
FS	finish-to-start
FF	finish-to-finish
SS	start-to-start
SF	start-to-finish
m	month
w	week
d	day
+	lag
–	lead

Creating a Network Diagram

A network diagram is a visual depiction of your schedule from beginning to end. It shows the tasks as boxes (also called nodes), linked by arrows. Once you are comfortable with the flow, or the diagram starts to get too big, you can switch from sticky notes to entering the information into scheduling software. Figure 7-2 shows the same information from Figure 7-1, but it is entered into software, and it shows the 10-day lag.

Network diagram: A visual depiction of the schedule using nodes and arrows.

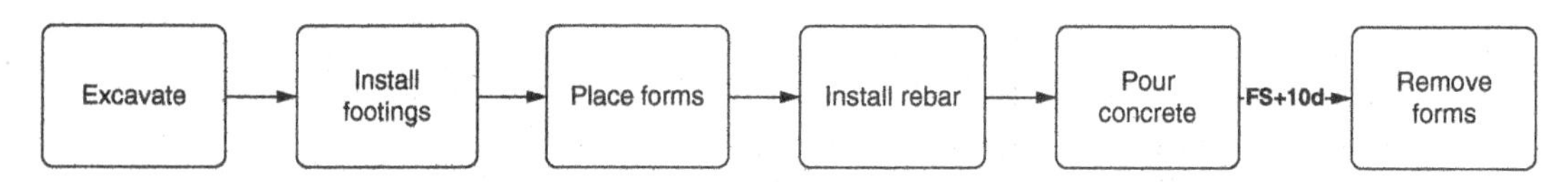

FIGURE 7-2 Network diagram in software.

Assign Team Members

You can have a perfect network diagram that shows the flow of work, but nothing happens without team members. At the start of the project, you will think about the roles and skills you need. As you progress, you will substitute names of team members for the roles. Entering the team member roles or names into scheduling software is referred to as resource loading.

Resource loading: Entering resources into a scheduling tool.

For large projects with 50 or more team members, you will need to spend some time thinking about the number of team members you will need for each role. In the next few sections you will learn about several easy tools you can use to help keep track of your team needs.

Resource breakdown structure: A hierarchy chart that shows resources grouped by type or category.

Resource Breakdown Structure

A resource breakdown structure can show the number of people you need in each role. The resource breakdown structure in Figure 7-3 shows an example for the renovation projects for the Dionysus Winery.

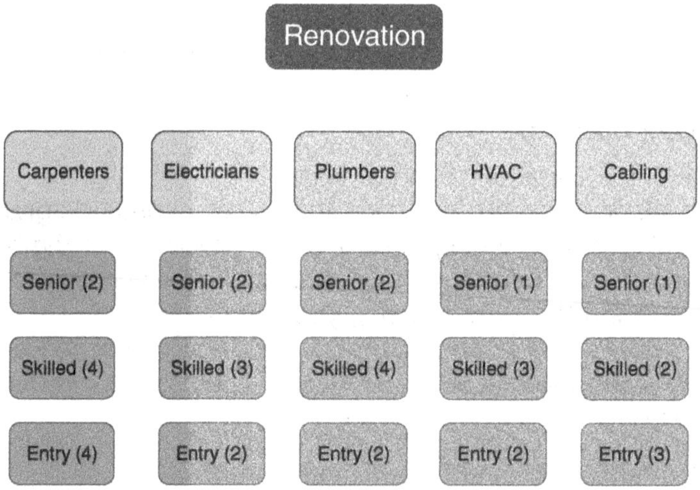

FIGURE 7-3 Resource breakdown structure.

This breakdown chart is like an org chart or a work breakdown structure, but it focuses on project resources. If you have a lot of physical resources, such as equipment, material, and supplies, you can either integrate those categories or do a separate breakdown structure for physical resources.

Resource Histograms

You can also show resource needs with a resource histogram. Figure 7-4 shows a resource histogram with the same information as the resource breakdown structure in Figure 7-3.

Resource histogram: A bar chart that shows information about resources, such as number or skills.

You can create a resource histogram by creating a table like Table 7-3 in a spreadsheet.

Then you simply highlight the data and then insert a column chart.

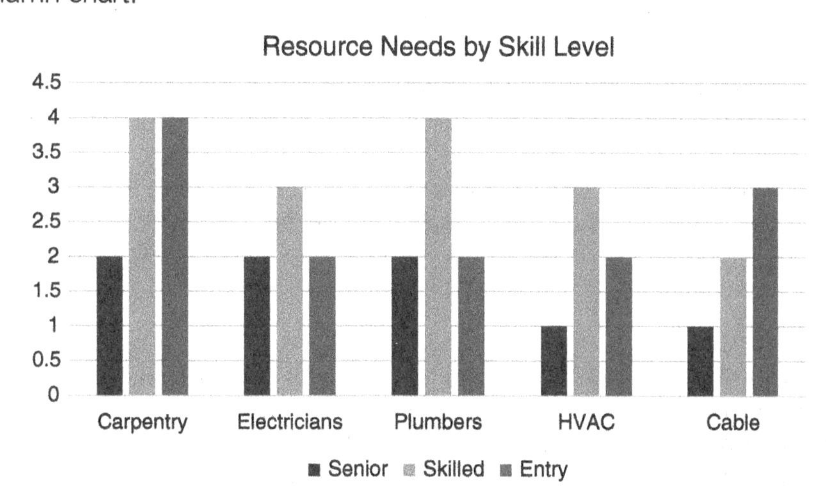

FIGURE 7-4 Resource histogram with resource needs by skill level.

TABLE 7-3 Resource Chart

	Senior	Skilled	Entry
Carpentry	2	4	4
Electricians	2	3	2
Plumbers	2	4	2
HVAC	1	3	2
Cable	1	2	3

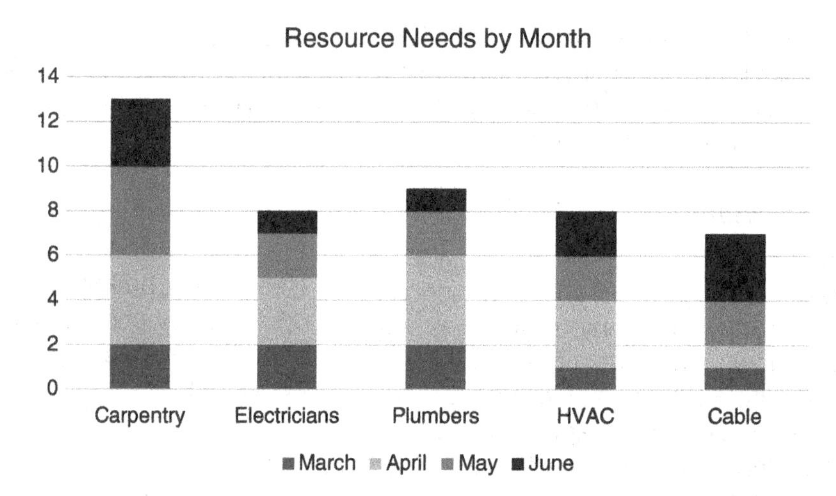

FIGURE 7-5 Resource histogram with resource needs by month.

Another great use of a resource histogram is to show the number of resources you will need by skill level, by month, or in each location. Figure 7-5 shows a resource histogram with resource needs by month.

This resource histogram shows a stacked bar chart rather than a column chart. There is no single correct way to chart your resource needs—use whichever method works best for you.

Role Descriptions

Many of the misunderstandings that occur on projects are a result of not having a clear understanding of the roles, competencies, responsibilities, and authority associated with each role. Having a document that clearly identifies this information can help minimize these types of misunderstandings. For example, you might record information like this:

Role: The position on the team. It describes the type of work that needs to be done, such as writer, programmer, or electrician. A role usually coincides with a job title.

Competency: Describes the skill or skill level of an individual. A junior plumber has the competencies to lay pipe and connect pipe following a set of blueprints. A senior plumber can design the plumbing to comply with standards and regulations.

Responsibility: The work the person in a role is expected to perform. The person with the role of an editor is responsible for reviewing all written content, ensuring that it conforms to the editorial guidelines and is easy to understand.

Authority: The right to make and approve decisions. The person in the role of the contract administrator is usually the only person with the authority to enter into or change a contract.

Responsibility Assignment Matrix

A responsibility assignment matrix (RAM) is a matrix chart that shows the type of participation for each role (or team member) and each work package. Like the role descriptions, this chart provides clarity around expectations for each team member and reduces the likelihood of misunderstandings.

A common version of a responsibility assignment matrix is called a RACI chart because it documents who is **r**esponsible, **a**ccountable, **c**onsulted, and **i**nformed.

- The *responsible* person is the person doing the actual work. This may or may not be the same as the accountable person.
- The *accountable* person is the one answerable for a deliverable.
- Someone who is *consulted* may be a person who has done this type of work before, a stakeholder with an interest in the outcome, the customer, or even a consultant.
- The person who is *informed* needs to know the status of the deliverable. They may have a dependent deliverable, or it may be the project manager or another interested stakeholder.

Role: A position on a team that describes the type of work the person does.

Competency: The skill level of an individual.

Responsibility: The work a person in a role is expected to perform.

Authority: The right to make and approve decisions.

Responsibility assignment matrix (RAM): A chart that shows the role team members have on a work package.

RACI chart: A type of responsibility assignment matrix that identifies who is responsible, accountable, consulted, and informed.

Table 7-4 shows a portion of a RACI chart using work packages and roles.

TABLE 7-4 Responsibility Assignment Matrix

	General contractor	Architect	Project manager	Cement contractor	Enculture manager	Production manager
Requirements	I	I	I		R	A/R
Drawings	C	A/R	C		C	C
Excavation	A			R		
Foundation	A			R		

With a RACI chart it is important to note that each deliverable must have one and only one accountable person since there should only be one person who is answerable for the status and success of a deliverable.

A RACI chart is a very common version of a RAM, but it is certainly not the only one. It can be useful to include an "S" for sign-off as well. If a sign-off function was being used in the RAM above, the project manager might sign off on the requirements the production manager submitted and the drawings the architect submitted.

Selecting Team Members

You may not have the luxury of choosing your team members. If you do have input into selecting team members, there are several variables to consider. Even if you don't get to choose who is on your team, keep these variables in mind so you can optimize team performance.

- **Skill level and experience:** A team member with a lot of experience and a high degree of skill will be able to complete the work in less time, with less rework, and with a higher degree of quality than someone with less experience. However, they often have a higher hourly pay rate.
- **Availability:** Team members who can dedicate themselves to working on the project full time will be more productive than those who can only work on it part time. The more projects a person is working on at the same time, the less productive they are on any one project.
- **Cost:** A project that uses in-house resources may not have to worry about the cost of team members, but if you are using contractors, you will want to consider the hourly or daily rate of team members.
- **Attitude:** Attitude can affect your project in a few ways. For example, a team member who really wants to work on the project may be a better fit for the team than someone with more experience who isn't excited about your project. Another aspect of attitude has to do with a person's general outlook. A person with a positive outlook is often a better team member than someone who is overly critical, combative, or negative.
- **Flexible skill set:** In Chapter 3 we talked about team members who are generalizing specialists, also known as T-shaped people. Team members with a deep understanding of a specific skill and a broad understanding of many complementary skills can be a valuable asset to the team. They can support other team members if they fall behind and present ideas from a holistic perspective rather than from a narrow focus.
- **Geography:** If you have a project that requires the team to meet in person, geography can be a significant factor in assigning team members. However, if you have

a partially or fully virtual team, you may be able to find team members with great skills and better rates than local talent. Keep in mind when establishing a virtual team, spending extra time on communications is critical. You want people to feel connected even though they don't get to see each other. In addition, when you can't read someone's body language, you miss some good communication cues, so establish as many communication channels as necessary to reduce the opportunities for misunderstandings is important.

Estimate Durations

Estimating the duration for the project, deliverables, or tasks takes place throughout the project—it is not a one-time event. Techniques for estimating duration and effort are described in Chapter 10: "Estimating." In this section we will differentiate between duration and effort and describe some of the variables that influence duration.

As you find out more information about the project, your estimates will become progressively more detailed and more accurate. At the beginning of the project you only have high-level estimates, such as milestones in the project charter. Once you start building your schedule, you can get a reasonably good idea of the durations for deliverables. Then using progressive elaboration techniques, you can start to refine estimates even more by calculating not only the duration but the effort involved as well.

> **Duration:** The amount of time needed to complete work.
>
> **Effort:** The amount of labor needed to complete work.

To understand the difference between effort and duration, assume you have a 40-hour workweek and a task that requires 80 hours of effort.

- If you have two team members who are working full time, you can assume that it will take 5 days, or 1 business week, to accomplish the work. The calculation would be 80 hours/2 = 40 hours of duration.
- If you have only 1 resource working at 50%, it will take 20 days, or 1 business month, to get the work done. 80 hours/.5 = 160 hours of duration.

As noted in the previous section on selecting team members, variables such as skill set, experience, and availability can affect the effort required and the duration of tasks.

If you want to get really accurate, factor in the assumption that for every 40 hours spent at work, people are productive only about 34 hours. The rest of the time they are answering emails, participating in staff meetings, or engaged in other non-project work.

Tip: Only spend as much time as necessary refining your estimates. If you don't need to track hourly or daily costs, calculating the effort hours is often not necessary.

At this point you will have an initial version of the schedule. However, your schedule is not ready for distribution yet. You will need to analyze it and likely find ways to compress it before you can baseline and share the schedule. We'll cover those topics in the next chapter.

SUMMARY

This chapter covers information on how to build a predictive schedule. The schedule management plan describes the scheduling methodology, tools, and how predictive and adaptive aspects of the schedule will be combined into an integrated master schedule.

We also looked at four steps used to build a predictive schedule:

1. Identify tasks;
2. Sequence tasks;
3. Assign team members; and
4. Estimate durations.

You start by decomposing work packages into tasks and then sequencing those tasks to create a network diagram. Once team members are assigned, they can help estimate task durations, and effort if necessary.

Key Terms

authority
competency
critical path
critical path methodology
duration
effort
finish-to-finish
finish-to-start
integrated master schedule
iteration
lag
lead
network diagram

RACI chart
release
release plan
resource breakdown structure
resource histogram
resource loading
responsibility
responsibility assignment matrix (RAM)
role
schedule management plan
start-to-start
story points
summary task
task
task board

8

Analyzing and Finalizing a Predictive Schedule

Developing your initial schedule takes some work, especially for a hybrid project. But once you have that initial schedule, you still aren't done. You will need to analyze the schedule, assess your resource allocation, and identify the critical path. Then you will often need to find a way to reduce the amount of time it takes to accomplish all the work. Only then can you create a schedule baseline.

In this chapter we'll discuss different aspects of analyzing the schedule to make sure it is optimized for your project. Then we'll look at two ways to compress the overall duration of the schedule. Finally, we will look at using buffer to protect your due date, and baselining the schedule.

ANALYZING THE SCHEDULE

Once you have developed the network diagram, loaded your resources, and estimated your durations, it is time to analyze the schedule. Your goal with this step is to determine if the schedule meets the needs of the stakeholders and the project constraints. There are four areas you should analyze:

- Convergence and divergence;
- Resource allocation;
- The critical path; and
- Float.

Convergence and Divergence

Convergence is when you have multiple tasks converging into one task; some people also refer to this as a sink. The point of convergence carries some risk with it because if any one of the paths finishes late, the convergence task starts late.

Convergence: A point in the schedule where multiple paths merge.

Divergence: A point in a schedule where one task separates into multiple paths.

Divergence is when multiple tasks are coming out of one task. This is sometimes referred to as a burst. If that one task finishes late, then all the diverging tasks start late. Figure 8-1 shows a network diagram with convergence on the left and divergence on the right.

Figure 8-2 shows the same information as Figure 8-1 but in a Gantt chart view.

Gantt chart: A bar chart used in scheduling where tasks are listed in rows and bars indicate the duration of the tasks.

It is a good practice to eliminate, or at least minimize, the number of convergence and divergence points in your schedule.

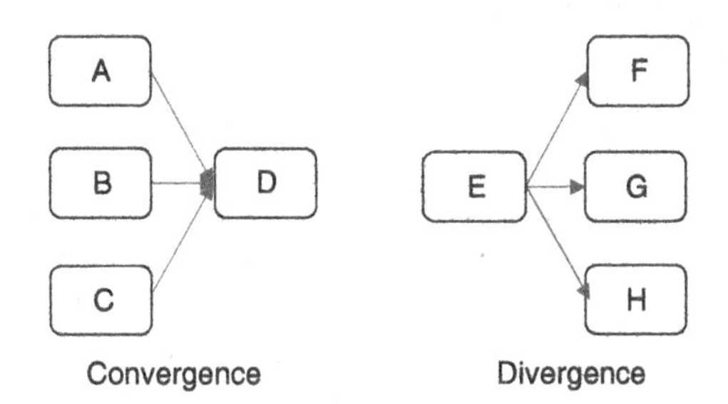

FIGURE 8-1 Network diagram view of convergence and divergence.

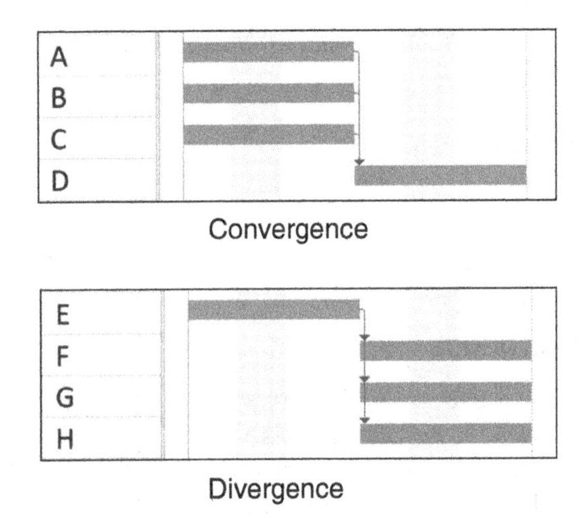

FIGURE 8-2 Gantt chart view of convergence and divergence.

Resource Allocation

Initially when you load your team resources into the schedule tool, you will likely find that some of your resources are overallocated—meaning the schedule shows they are working more than their allocated time. For team members who are working on your project full time, this means they are scheduled for more than 40 hours of work per week. If you are only tracking duration and not effort, this may not be a problem. For example, if a team member is scheduled to work on three tasks at the same time, and each task only requires 20 hours of effort, and the duration shows two weeks, then the person is not overallocated. However, if each task requires 40 hours of effort, then they are overallocated.

For situations when a team member is overallocated, you can apply resource smoothing or resource leveling. Resource smoothing adjusts activities within their float amounts but does not change the critical path. In other words, some resources may remain overallocated.

Resource smoothing: Reallocating resources within the available float, so the critical path is not affected.

Resource leveling: Reallocating resources so no instances of overallocation remain.

Resource leveling adjusts task dates so that resources are no longer overallocated. Leveling frequently extends the due date. As shown in Figure 8-3, if a resource is scheduled to work on task A and task B during the week of March 1 through March 5, and both tasks require five days of full-time commitment, resource leveling

Before Leveling

	1	2	3	4	5
Task A	8	8	8	8	8
Task B	8	8	8	8	8
Total Hours	16	16	16	16	16

After Leveling

	1	2	3	4	5	6	7	8	9	10	11	12
Task A	8	8	8	8	8							
Task B								8	8	8	8	8
Total Hours	8	8	8	8	8			8	8	8	8	8

FIGURE 8-3 Leveling resources.

would extend the start date of task B to begin on March 8 and end on March 12. This way your resource is not overallocated.

Figure 8-4 shows another type of resource leveling where the goal is to reduce peaks and valleys in resource utilization. For example, instead of working 40 hours one week and 10 the next, resource leveling seeks to have a resource work 25 hours each week.

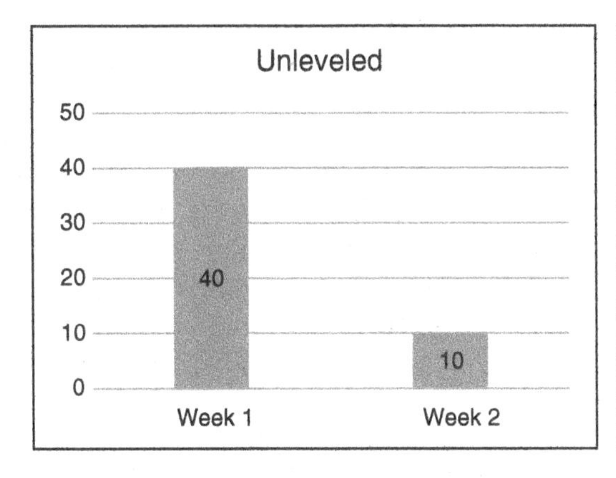

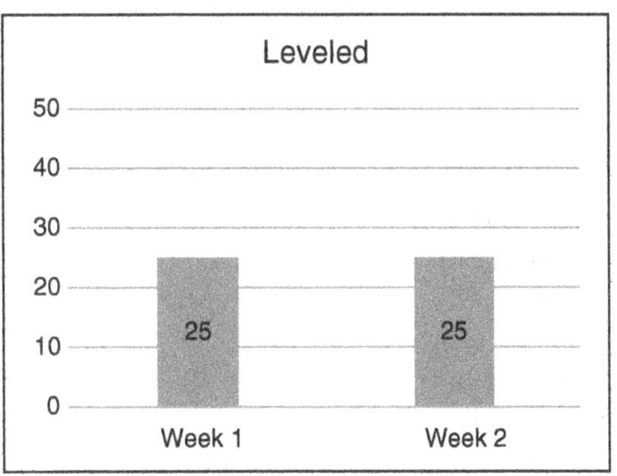

FIGURE 8-4 Leveling hours.

The Critical Path

Recall the critical path is the series of tasks through a project with the longest duration, which determines the soonest the project can be completed. You will definitely need to spend some time analyzing work on the critical path. Five areas you should focus on are:

1. Are there any tasks with high risks on the critical path? If so, see if there is a way you can rearrange the network so tasks with high risk are not on the critical path.
2. Is it logical? The critical path shouldn't be determined by support type of work, such as project management. It should be determined by work with defined deliverables.
3. Are there any near-critical paths? If work on any of the near-critical paths is late, that path could become the new critical path.
4. See if there are any sinks or bursts (convergence and divergence points) on the critical path. Where possible, eliminate or reduce those as they add risk to the path.
5. Determine if there are areas where you should add buffer. We discuss buffer in the section on finalizing the schedule.

Float

Float occurs on tasks that are not on the critical path. Float gives you flexibility and room to optimize your schedule. There are two types of float: total float and free float. Total float is the amount of time a task can slip without affecting either a project constraint or the end date. This is what most people refer to as float.

> **Total float:** The amount of time a task can slip and not affect a project constraint or the end date.
>
> **Free float:** The amount of time a task can slip and not affect the following task.

Free float is the amount of time a task can slip and not affect the following task. Free float is found when two or more paths converge. Assuming one path has a longer duration, free float will be on the last task of paths with shorter duration. Those are the only tasks that can slip and not affect the following task.

Figure 8-5 shows a network diagram with total float and free float. The letters represent the task name and the numbers the task duration in days. Thus, A2 is task A, which is two days in duration.

You can see that by adding up the duration in tasks A, B, and C you have a total duration of 13 days. Adding up the path with D and E gives you a total duration of 18 days. Therefore, the critical path is D and E. The path of A, B, and C has 5 days of total float.

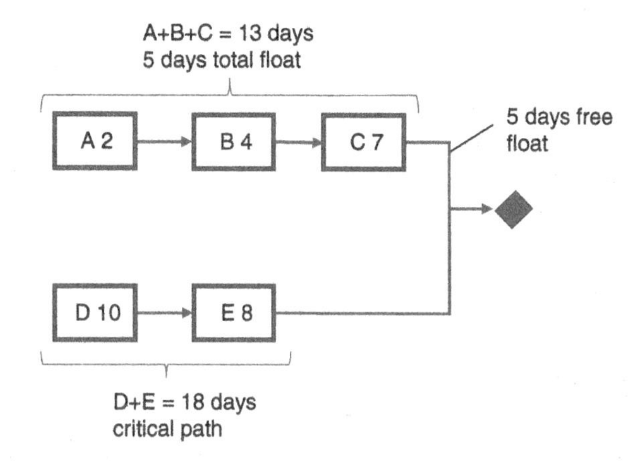

FIGURE 8-5 Network diagram view of total float and free float.

In other words, that path can absorb 5 days of slippage before it affects a constraint (the milestone where the paths converge is a constraint). This demonstrates that total float is shared by all the tasks on the path. Any float used by A reduces the float available to tasks B and C. Only task C has free float. Assuming A and B don't use any float, then C can slip by up to 5 days and not affect any other task on the network.

Figure 8-6 shows the same information in a Gantt chart.

Free float is very useful if you have resources that are only available on the days they are scheduled. If the resources for B and C are scheduled to work only on the days they are scheduled and A takes longer than the two days it is scheduled, then there is a problem because the resources for B and C may not be available, which could potentially cause the project to fall behind while you wait for a resource that is available to do the work for tasks B and C. Therefore, when you are analyzing float, you should identify any areas that have free float, as they have the most schedule flexibility.

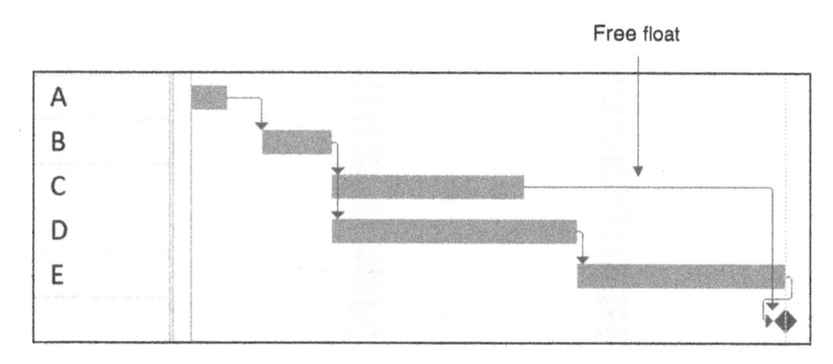

FIGURE 8-6 Gantt chart view of total float and free float.

You should also look at the amount of float on various paths. If you have any tasks with a large amount of float, double-check to make sure you didn't miss a dependency. There may be a legitimate reason for having a lot of float; however, it is not very common, so make sure to review those paths with a significant amount of float.

If you do find paths with a good degree of float, you may want to see if some of the resources assigned to those tasks can help with work on the critical path. This may not be possible if the team members on the paths with float don't have the skill set necessary to do the work on the critical path, but it is a good idea to check.

FINALIZING THE SCHEDULE

Finalizing the schedule can include looking for ways to compress the schedule to meet a target delivery date, identifying areas where you should include some buffer, and finally, baselining the schedule.

Schedule Compression

Compressing a schedule involves looking for ways to shorten the overall project duration without reducing the project scope. There are several ways to compress the schedule. The most obvious is to look for places you can reduce bureaucracy or find more efficient ways to work. This usually involves eliminating non-value adding tasks. You can also look for ways to compress the schedule by crashing or fast-tracking. These methods are described in the next two sections.

Schedule compression: Reducing the schedule duration without reducing scope.

Crashing the Schedule

Crashing: Compressing the schedule by adding resources, such as people or money.

Crashing the schedule involves cost/schedule trade-offs. In other words, you look for ways to shorten the schedule by applying more resources or by spending more. The intent is to get the most schedule compression for the least amount of money. Some common ways of accomplishing this include:

- **Bringing in more resources:** Sometimes, more people working, or using additional equipment, can speed up progress. Be careful. Sometimes adding resources extends the duration because coordination, communication, and conflict take more time than they save!

- **Working overtime:** Sometimes, staff can work longer hours or work weekends. However, this is useful only for a few weeks. After that, people burn out and are less productive, so this is a short-term fix. You also need to take into consideration any union or labor regulations.
- **Paying to expedite deliverables:** This solution can include overnight shipping and paying bonuses to contractors for early delivery.

You won't be able to crash everything. For example, you can't expedite the occupancy permit inspection by paying for two inspectors, and you can't pay them to expedite the process. Because crashing usually increases spending, you will want to weigh the benefit of earlier delivery against the cost.

Fast-Tracking the Schedule

Fast-tracking shortens the schedule by overlapping activities that are normally done in parallel. One way of doing this is changing the network logic by using leads and lags. For example, you can change a finish-to-start relationship to a finish-to-start with a lead. This causes the successor task to start before the predecessor is complete. You can also change the relationship to a start-to-start with a lag or finish-to-finish with a lag.

> **Fast-tracking:** Compressing the schedule by performing work in parallel that is usually performed sequentially.

Here are a few examples from the hotel construction at the Dionysus Winery:

- If the original construction schedule was set up with finish-to-start dependencies among framing, electrical, plumbing, and HVAC, you can probably conduct some of the electrical, plumbing, and HVAC in parallel instead of sequentially.
- If you have a finish-to-start relationship with the rough plumbing (running pipes to the building and rooms) and the finish plumbing (installing sinks, toilets, showers, fixtures, etc.), you can change the relationship from finish-to-start relationship to start-to-start with a lag. For example, after half of the bathrooms have all the pipes, the fixtures can start to be installed. The finish work doesn't have to wait until all the bathrooms, the common areas, and kitchens plumbing is roughed in.

Fast-tracking can increase your risk on the project and might necessitate rework. Make sure you understand the real relationship between activities when you fast-track, or else you will end up with a schedule that doesn't make sense and can't be executed.

Schedule Buffer

Even if you do an excellent job of planning your work and you have highly skilled resources, sometimes events don't go as planned. One way to protect your delivery dates is to include buffer in your schedule. Buffer is extra time where no work is scheduled. Before you finalize your schedule, you may want to revisit those areas that have the most risk, such as points of convergence and divergence, phase gates, key deliverables, paths that feed into the critical path, contractually obligated dates, or final delivery. These are all points in time where being late is unacceptable.

> **Buffer:** Time inserted into the schedule to protect delivery dates.
>
> **Feeder buffer:** Time inserted into the schedule where a path merges with the critical path.
>
> **Project buffer:** Time inserted into the schedule before the final delivery date.

Feeder buffer is used when you have a path that is merging into the critical path. To insert feeder buffer, put a block of time after the last task in the path feeding into the critical path. This increases the chances that if the feeder path is running late, it won't affect the critical path.

Project buffer is used between the scheduled end date and the promised delivery date. Therefore, if anything happens on the project that negatively affects the project duration, you are protected from delivering the project late.

Figure 8-7 shows a Gantt chart with feeder buffer and project buffer.

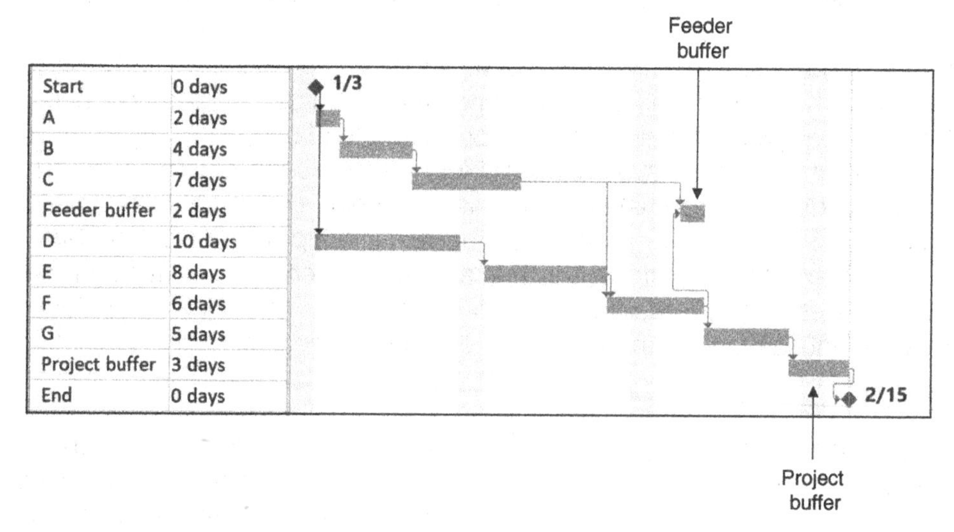

FIGURE 8-7 Gantt chart with buffers.

For the feeder buffer I inserted a start-to-start relationship with a lead so that the feeder buffer always starts two days before it merges with the critical path. That way, if the critical path changes, so does the feeder buffer.

Tip: Buffer is not the same as float, and it is not the same as padding. Buffer is the deliberate insertion of time to protect key delivery dates. Float is the amount of time a task can slip without making the schedule late. Float is based on the network diagram. Padding is adding time to the overall project without taking time to analyze the schedule. It lacks the discipline of assessing the schedule and strategically adding time to produce better outcomes.

Baselining the Schedule

A schedule baseline is an agreed-upon version of the schedule that will be used to measure progress and detect variance. Once you have analyzed your schedule, resolved any resource allocation issues, assessed your critical path and float, and added any buffer you feel is appropriate, you can share your schedule with key stakeholders and gain their agreement for baselining.

Baseline: An agreed-upon version of a document that is used to measure progress and detect variance.

Baselining the schedule and the budget usually occurs at the end of a planning phase for predictive projects. Once the baseline is approved, it becomes the benchmark that you will use to measure progress for the rest of the project. The baseline should only change due to the following circumstances:

1. A risk or issue occurs that disrupts the agreed-upon baseline;
2. The performance variance is so great that measuring against the baseline no longer provides value; and
3. An agreed-upon change is incorporated into the project necessitating a new baseline.

The schedule baseline is usually kept under change control or configuration control. This means that the baseline can only be changed under the three circumstances indicated above and that it can only be changed with approval.

Some project managers keep the baseline schedule at a high level and a working schedule at a more detailed level. The working schedule has to have the high level as the baseline but provides more detail and flexibility for the day-to-day management of the project.

SUMMARY

This chapter covered information on analyzing and finalizing a predictive schedule. Analyzing the schedule consists of addressing areas of convergence, divergence, and resource allocation. You also assess the critical path and the float. To finalize and baseline the schedule, you may need to look for ways to compress the duration, such as crashing and fast-tracking. The final steps are assessing if you need buffer, and then creating a schedule baseline.

Key Terms

baseline
buffer
convergence
crashing
divergence
effort
fast-tracking
feeder buffer
free float
Gantt chart
project buffer
resource leveling
resource smoothing
schedule compression
total float

9

Adaptive and Hybrid Scheduling

Scheduling works differently with adaptive environments than with predictive environments. As you saw in the previous two chapters, in a predictive environment the goal of a schedule is to organize the work sequentially. This process allows you to determine when your team will be working on various deliverables and when the project will be complete. When working with an Agile methodology, the team determines how much scope can be accomplished in a fixed duration work period. The goal is to provide value early and often, while staying open to shifting priorities.

In this chapter we will look at release and iteration planning to provide a high-level plan for creating and delivering value. Then we'll look at how task boards track the detailed work. We'll finish the chapter by looking at a few ways you can blend predictive and adaptive scheduling for a hybrid project.

ADAPTIVE SCHEDULING

When you are working with deliverables where the scope evolves and changes based on customer, market, and stakeholder feedback, at the start of the project you don't know all the work you will need. Often you don't even know all the work you will need when you are mid-way through the project. Thus, scheduling in this environment requires progressive elaboration of the scope and therefore rolling wave planning for scheduling the deliverables.

There are three types of adaptive work: incremental, iterative, and Agile.

- Incremental development work begins with a simple deliverable and then progressively adds features and functions;
- Iterative development work begins with delivering something simple and then adapts based on input and feedback;
- Agile is an adaptive way of delivering value by following the four values and 12 principles established in the Agile Manifesto.

Iteration: A brief, set time interval in a project where the team performs work. Also known as a timebox or sprint.

Projects that use Agile schedule work in iterations that are a set duration, such as one week, two weeks, or four weeks. Iterative and incremental development approaches may or may not use fixed duration timeboxes for their iterations, depending on the nature of the deliverables. Software development usually uses fixed duration timeboxes, though other iterative and incremental projects may not. The regular cadence of fixed iterations encourages synchronization and predictability.

Release Planning

In an adaptive environment the team develops a release plan that provides a high-level view of when certain features and functions will be available for use. Of course, this may change if the priorities of deliverables change, but it provides a loose plan to follow.

Release: A set of features, functions, or deliverables that are placed into use.

Release plan: A plan that shows the expected timing, milestones, and outcomes for releases.

A release plan starts with the project or product roadmap. We'll use the system development deliverable for the Dionysus Winery to demonstrate how a release plan is developed. The roadmap for the Dionysus Winery (shown in Figure 5-4), shows that the system development will occur starting partway through month 6 and will finish at the end of month 11. Angelo Romero, the operations manager, will serve as the product owner for the system. He has indicated the priority for the various functions as:

1. Inventory management;
2. Winery management (making and aging wine);

3. Vineyard management;
4. Wine club membership; and
5. Forecasting.

Remember: A minimum viable product is the first release of a product that contains the least number of features or functions in order to be useful.

The minimum viable product for the system is the inventory management function; therefore, that will be the first release.

Angelo and Tony Dakota (the project manager) decide to ask Sophie Rojas to be the scrum master for the system development work. The three of them looked at the prioritized functionality and estimated when the various functions could be developed, tested, and released. After some discussion they came up with the release plan in Figure 9-1.

Sophie will work with the system development team to plan release 1 in more detail. Releases 2–4 will stay at a high level as the team learns more about stakeholder needs and how users will interact with the various system applications.

The development team decides to work in two-week iterations. They will use the first two weeks (often called sprint, or iteration 0) to get set up. Set-up can include arranging their working space, making sure the team has all the equipment and supplies they need, determining the processes they will use to develop and test, and other team and logistical issues. They also review the work they will be doing in the first release and develop an iteration plan (also known as a sprint plan) that identifies the work they expect to do in each iteration. Figure 9-2 shows the team's iteration plan.

For every iteration except iteration 0, the team will hold an iteration planning meeting with the product owner to review the work on the prioritized backlog. They will ask questions to get clarification around the backlog items and define the acceptance criteria. By the end of this meeting the team will have estimates and a plan for the work in the current iteration.

Iteration planning: A meeting used to clarify and estimate the work that will be accomplished in the current timebox. Also known as sprint planning.

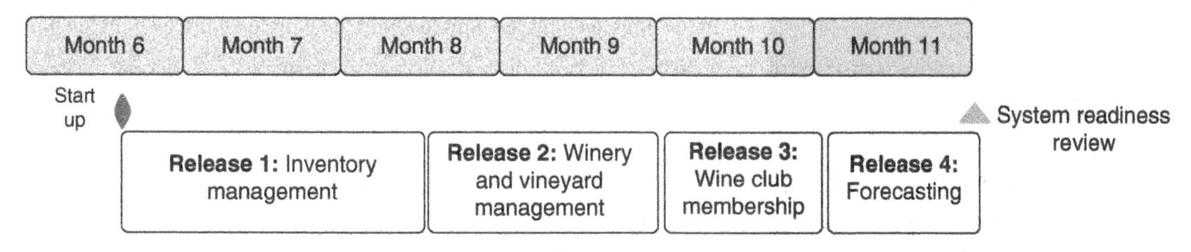

FIGURE 9-1 Dionysus Winery system development release plan.

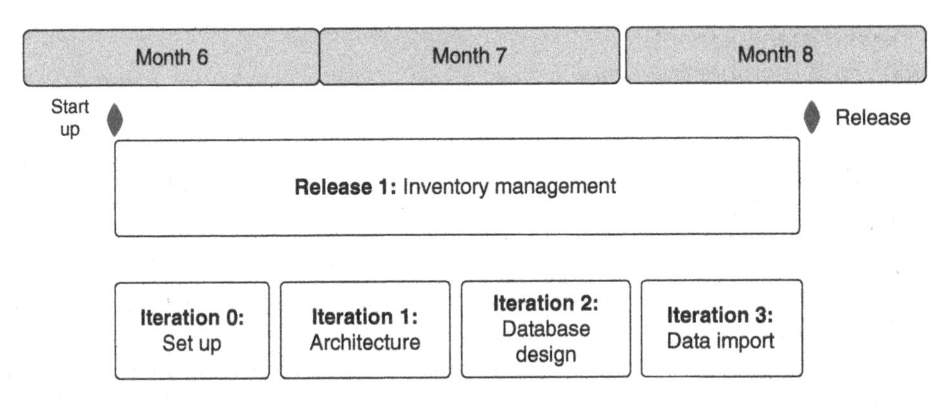

FIGURE 9-2 Winery system development iteration plan.

Retrospective: A workshop to review the work product and processes to find ways to improve outcomes.

At the end of the iteration, the team will demonstrate the functionality to Angelo, Tony, and other relevant stakeholders. The stakeholders will accept the work and/or provide feedback. After the demonstration the team will hold a retrospective to consider their work process, determine if there is anything they want to change, and decide if there is anything new they want to try in the next iteration.

Task Boards

When an iteration begins, the work for the iteration is posted on a task board. A task board has multiple columns that indicate the status of work. They can also be called kanban boards (kanban is the Japanese word for "sign"), scrum boards, Agile boards, and other similar names. Task boards can be electronic (in the cloud or on a server) or low tech with whiteboards and flipcharts. For teams that are in the same location, low tech is preferred. The simplest task boards, such as the one shown in Figure 9-3, have columns for "To Do," "Doing," and "Done."

Task board: A display of project work that allows stakeholders to see the current status of tasks.

Teams can decide what makes sense for tailoring their task boards. Some examples are listed below.

1. Add a column for "Testing" or "Verify";
2. Substitute "User Stories" for "To Do";
3. Substitute "Backlog" for "To Do."

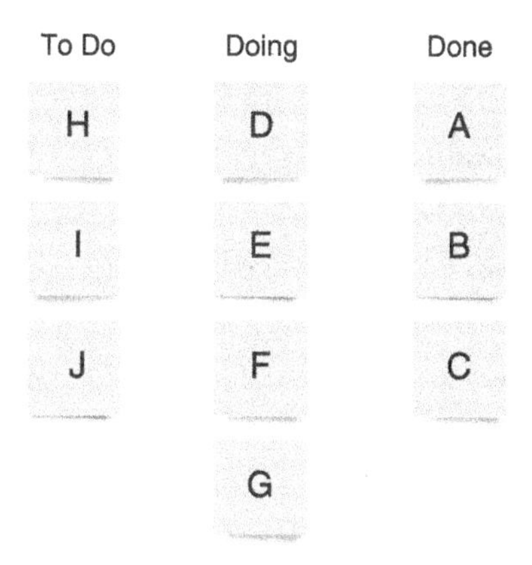

FIGURE 9-3 Sample task board.

Some project teams employ a task board but don't use iterations. This is called flow-based scheduling. In flow-based scheduling, when a team member is available, they work on the next prioritized task on the backlog. This keeps the work flowing continually. For flow-based scheduling you may see a person in the role of a flow master or solution delivery manager rather than a scrum master.

Even if a team is not using iterations, they should still meet at least monthly to reflect on performance and metrics and incorporate feedback for planning and improvement purposes. Flow-based scheduling is harder to represent on an integrated master schedule, but it can be an effective way to accomplish work.

HYBRID SCHEDULING

Hybrid scheduling is a blend of predictive methods (as described in Chapters 7 and 8) and adaptive methods to create an integrated master schedule. There are several ways you can schedule a hybrid project. We cover three of them in the next sections, but don't feel limited to just these three methods; the great thing about hybrid projects is you can tailor them to meet your needs.

Predictive with Releases and Iterations

This type of hybrid plan uses a predictive schedule and creates a placeholder for releases. The detailed work for each iteration is managed with task board software or whiteboards if the team is in the same location. Figure 9-4 shows an example of this option where the predictive work is shown as summary tasks, and the adaptive work is shown at the release level.

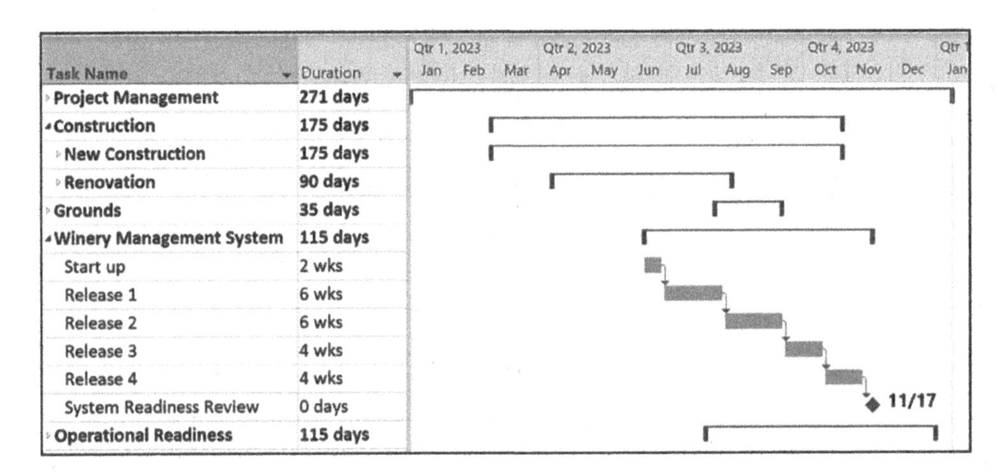

FIGURE 9-4 Predictive schedule with releases.

Predictive with Iterations Inserted

Some scheduling software allows you to build a predictive schedule or an adaptive schedule. You can build each type of schedule separately and insert the adaptive schedule into the predictive schedule. Figure 9-5 shows an example of this option. The predictive work is shown at a summary level. The adaptive work shows all the releases with the first release elaborated to show the work for each iteration.

Adaptive then Predictive

New product development projects often use an adaptive approach to identify and develop the features and functions for the product. Once the product is finalized, the project moves to a predictive approach for manufacture, packaging, marketing, and distribution.

Dependencies in Hybrid Schedules

One of the ways hybrid projects get into trouble is balancing the structure needed for a predictive schedule with the evolving nature of adaptive scope. When a project component is developed using adaptive measures, the team developing that component needs to be aware that they are part of a larger project. There are dependencies between the adaptive work and the predictive work that must be honored for the project to deliver on time.

In the Dionysus Winery project, the wine club management component must be released before the wine club marketing campaign begins. As members register for the wine club their information will be entered into the wine club management application.

Task Name	Duration
Project Management	**271 days**
Construction	**175 days**
New Construction	**175 days**
Renovation	**90 days**
Grounds	**35 days**
Operational Readiness	**115 days**
Hybrid PM Sprints	**170 days**
Inventory Management Release	**8 wks**
Set-Up Iteration	**2 wks**
Establish ways of working	2 days
Acquire equipment	3 days
Set up environment	5 days
Architecture Iteration	**2 wks**
Review enterprise architecture	10 hrs
Review enterprise architecture policies	10 hrs
Identify interfaces	20 hrs
Develop process diagram	30 hrs
Database Design Iteration	**2 wks**
Identify data sets	10 hrs
Build data model	40 hrs
Build prototype	40 hrs
Data Import Iteration	**2 wks**
Import data	2 days
Clean data	5 days
Write user instructions	3 days
Winery and Vinyard Management Release	**6 wks**
Wine Club Membership Release	**4 wks**
Forecasting Release	**4 wks**

FIGURE 9-5 Predictive schedule with adaptive schedule inserted.

Additionally, the entire system must be completed and released in time for the staff to receive training before the grand opening.

These dependencies between the predictive and adaptive parts of the project necessitate ongoing communication between the development team (including the scrum master), product owner, and project manager. Therefore, particularly in hybrid projects, evolving scope doesn't mean ongoing development—there are still deadlines that must be met.

SUMMARY

In this chapter we discussed how adaptive work begins with a roadmap and then develops a release plan that identifies when various functions will be ready for use. Releases are composed of iterations or sprints, which are set durations where work is developed and delivered. The work in each iteration is prioritized in a backlog and is often managed on a task board.

Hybrid scheduling uses both predictive and adaptive scheduling. We described three ways to develop hybrid integrated master schedules. However, with hybrid projects, you can tailor your scheduling method to meet your needs.

Key Terms

iteration
iteration planning
release
release plan
retrospective
task board

10

Estimating

Estimating is one of the more challenging aspects of managing a project; it is also one of the most important. Project managers develop estimates for effort, duration, resources, costs, and even reserve. That is a lot of estimating. Estimating for a hybrid project is even more challenging. Different development and work methods use different estimating methods. Therefore, there are a number of different techniques you can use to estimate, and because of that, there are different ways to track progress and report on project status.

In this chapter we will talk about how estimating ranges and accuracy evolve over the project life cycle. We will describe six techniques for estimating and when each technique is best used. Then we will look at how you can use cost estimates to develop a project budget.

ESTIMATING RANGES

When you are first given a project, you usually have a broad understanding of the project objectives and the desired outcomes. In some cases, you have data from similar projects that can help you estimate the cost and duration. In other situations, there is no previous historical data that you can reference, which makes the estimates more uncertain. It stands to reason that the more you understand about the scope, available resources, the nature of the work, and the environment, the more accurate your estimates will be.

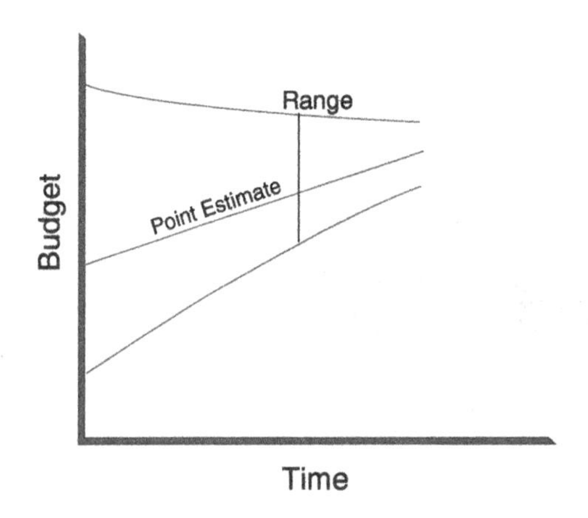

FIGURE 10-1 Range estimates over time.

Point estimate: A single value that represents the best prediction for an outcome.

To account for the uncertainty associated with estimating, we usually provide a range of estimates. At the beginning of a project, you are likely to have a larger range of estimates. For example, you may have a point estimate for the budget of $5,000,000, and a range of ±50%. This is especially true for projects that are developing new technology or using unproven processes.

As you learn more about the intended outcomes by gathering requirements, developing prototypes, and engaging in detailed planning, you can reduce the uncertainty. This may lead to a range that is ±25%. Once you have a resource loaded schedule, a risk register with risk responses, and have agreement on the scope and requirements, you may be able to reduce your range further to ±10%. Figure 10-1 shows how a cost estimate evolves over time with the range narrowing and the point estimate going up.

For components of the project with evolving scope the sponsor (or the person funding the project) may choose to keep the budget fixed and get the most functionality for the funds allocated, or they may decide to increase the budget to get more features and functions.

ESTIMATING METHODS

There is a saying about using the right tool for the right job—that goes for estimating as well. There are numerous methods for estimating but knowing *when* to use each method is as important as know *how* to use each method. Table 10-1 provides a brief description of each method and describes when it can best be used.

TABLE 10-1 Estimating Methods

Method	Description	When Used
Analogous	Using information from previous similar projects to develop an estimate for the current project.	Most often for predictive projects. Can be used to estimate effort, duration, cost, resources, and reserve.
Parametric	Developing a mathematical model based on significant historical data.	Most often for predictive projects. Usually used for cost but can be used for effort, duration, resources, and reserve.
Multipoint	Developing optimistic, pessimistic, and most likely estimates and calculating an average or weighted average.	Most often for predictive projects. Can be used to estimate effort, duration, cost, resources, and reserve.
Affinity grouping	Categorizing elements into groups based on similar characteristics.	Used for adaptive projects. Most often used to estimate effort, duration, or story points.
Wideband Delphi	Working with a group of experts to develop individual estimates, discuss, and reestimate until consensus is reached.	Used for adaptive projects. Most often used to estimate effort, duration, or story points.
Bottom-up	Summing estimates from individual work packages to arrive at an overall estimate.	Used for predictive projects. Most often used to estimate costs and aggregate them into a budget.

The following sections provide a more detailed description of each method and an example of their application.

Analogous Estimating

Analogous estimating is the most common method of estimating. In its most basic form, analogous estimating compares past projects with the current project, determines the areas of similarity and the areas of difference, and then develops an estimate based on that information.

Analogous estimating: Using information from previous similar projects to develop an estimate for the current project.

A more robust application determines the cost or duration drivers and analyzes the relationship between past similar projects with the current project. This can include size, complexity, risk, number of resources, weight, or whatever other aspects of the project influence your estimate.

Analogous Estimating Example

For this example, we will create an analogous estimate for developing course materials for a two-day orientation training for Dionysus Winery employees.

The HR specialist who will be developing the materials states that she developed a similar training program for a previous employer—but that training took three days instead of two. However, this class is a bit more complex, which means it takes longer to develop the materials. The previous class had four hands-on demonstrations as part of the class, while this one only has three.

The HR specialist checks her records and comes back with the following information.

The previous class was three days long and required 200 hours of development time. The content was relatively simple. There were four demonstrations that were part of the class. Setting up those demonstrations required an additional 40 hours.

To utilize this information to create an estimate for the winery orientation we need to apply a modifier to account for the differences. Table 10-2 shows the information organized in a table.

Because the previous class was two days long and this class is three days long, we modify the estimate by 33%. The HR specialist indicates that the increased complexity will entail about 10% more time to create and check the content. With four demonstrations that took 40 hours to create, we can assume 25% less time for the demonstrations. Applying the modifiers to the previous class results in the modified estimate shown in Table 10-3.

TABLE 10-2 Analogous Estimate Part 1

Previous class	Effort (hours)	This class	Modifier
3 days	200	2 days	–33%
Easy	Included	Difficult	+10%
4 demos	40	3 demos	–25%
Total	240		

TABLE 10-3 Analogous Estimate Part 2

Previous class	Effort (hours)	This class	Modifier	Modified estimate (hours)
3 days	200	2 days	–33%	133
Easy	Included	Difficult	+10%	13
4 demos	40	3 demos	–25%	30
Total	240			176

Uses and Benefits

Analogous estimates are most commonly developed early in the project at a high level, with an expectation that the estimate will evolve and get increasingly detailed as more information is uncovered. The benefits of using analogous estimates are that they are relatively quick to develop and they're not very costly to develop. However, because they're usually done at a high level, they're not the most accurate method.

To use this method effectively, projects must be similar in fact, not just in appearance. A software upgrade may sound similar but upgrading a Windows application is not the same as updating a Word application.

Parametric Estimating

Parametric estimating uses a mathematical model to develop estimates. Not all work can be estimated this way, but when you can, it's fast and easy. You can see how to use parametric estimating for calculating effort and duration below.

Parametric estimating: Developing a mathematical model based on significant historical data.

- Assume the painting contractor for the winery is estimating how long it will take to paint the 26,000 square foot wine storage facility. He knows that on average it takes one hour to paint 100 square feet. He can divide the total square feet by the 100 to get the effort hours: 26,000/100 = 260 hours of effort.
- If the painting contractor has five people working on the job for eight hours per day, that equates to 40 hours of effort per day. Divide 260/40 = 6.5 days of duration.

Parametric estimating is often used in construction to get a high-level cost per square foot estimate. When estimating the cost to build the hotel, the general contractor stated the costs would be about $225 per square foot. If the hotel has 30,000 square feet, the estimated cost would be $6,750,000.

Multipoint Estimating

Multipoint estimating is an excellent method to use with work packages that have a lot of uncertainty, risk, or unknowns. This method provides an estimated range and an expected value.

To use this method, collect three estimates based on the following scenarios.

Multipoint estimating: Developing optimistic, pessimistic, and most likely estimates and calculating an average or weighted average.

- **Optimistic:** The optimistic (best-case) scenario means that everything goes as planned. For duration estimating it means you have all your required resources, nothing goes wrong, everything works the first time, and so on. For cost estimating it means there are no price escalations, no scrap or rework, or other events that could cause the cost to go up. This is represented as an *O,* for "optimistic."

- **Most likely:** The most likely scenario considers the realities of project life, such as someone being called away for an extended period, work interruptions, an increase in the cost of material, and so forth. This is represented as an *M,* for "most likely."
- **Pessimistic:** The pessimistic (worst-case) estimate assumes much rework, delays in work getting accomplished, unexpected price increases, and other issues. This is represented as a *P*, for "pessimistic."

The range is between the optimistic and pessimistic estimates. The reality is that the actual duration or cost will fall somewhere between those estimates, not on the extremes. One way of using the multipoint estimate is just to calculate the average of the three estimates. A better way is to develop a weighted average where you weight the most likely scenario more than either the optimistic or pessimistic estimates. The most common equation is known as a beta distribution: $\frac{O+4M+P}{6}$. This equation weights the most likely estimate four times more than the optimistic and pessimistic estimates. The following examples show the differences between the average and the weighted average.

Multipoint Estimating Examples

Assume the cement contractor is working on developing an estimate for putting in the foundation for the hotel, restaurant, and tasting room. He has experience working in the area and is familiar with the composition of the soil. He indicates that the best he can hope for is an estimate of $50,000. However, he did some work on another winery where part of the land was very unstable and had to be shored up, and other parts were sitting on granite and needed a lot more effort to get the site ready. That job cost $120,000. Based on a preliminary look around, he estimates that this job will be around $70,000.

Calculating the average, the estimate would be $\frac{50{,}000+70{,}000+120{,}000}{3} = \$80{,}000$.

If you do a weighted average, the estimate would be $\frac{50{,}000+4(70{,}000)+120{,}000}{6} = \$75{,}000$.

In this example you can see how the difference in the most likely and the pessimistic estimate has a bigger impact when you just average the numbers. However, that variance is reduced when you give a higher weight to the most likely estimate.

Uses and Benefits

A multipoint estimate is a great method to use when there is a lot of uncertainty. It determines a range of outcomes and identifies a point estimate. While the beta distribution is the most common equation used, you can tailor the weighting to reflect whatever you think is realistic. For example, if you think there is a higher probability that things will go wrong, you can weight the pessimistic estimate more. The drawback of using this method is that it can be hard enough just getting one estimate from people—asking them to provide three may be asking a bit too much.

Affinity Grouping

This method is used on adaptive projects. It is a quick way to group user stories into similar-size buckets. Rather than figuring out if a chunk of work will take 10 hours or 20 hours of effort, team members group work in categories that take approximately the same amount of work.

Affinity grouping: Categorizing elements into groups based on similar characteristics.

Some of the fun ways to group is by coffee cup size, dog size, t-shirt size, or the Fibonacci sequence.

Coffee cups	Short, Tall, Grande, Venti
Dog sizes	Yorkie, Spaniel, Labrador, Mastiff
T-shirts	XS, S, M, L, XL
Fibonacci	1, 2, 3, 5, 8, 13, 21, . . .

Fibonacci sequence: A series of numbers in which a number is the sum of the two preceding numbers.

One of the challenges with affinity grouping is that it is a form of relative estimating where each project team sets their own standard. There is no universal standard that 3 Fibonacci points = 15 hours of work, or a Labrador = 20 hours of work. The estimates are relative to the other work a specific team is doing.

Affinity Grouping Example

The release and iteration plans the team developed, shown in Figures 9-1 and 9-2, from the previous chapter can be decomposed into smaller chunks of work that can be estimated using affinity grouping. The team decides to use the Fibonacci sequence to estimate the work for iterations 1–3. Iteration 0 is set up, so they don't bother estimating that iteration. The team works with the product owner, Angelo Romero, to understand his needs. Once they understand what he wants, they estimate the work. Figure 10-2 shows the following:

1. The tasks needed to accomplish the work for each iteration in release 1;
2. The points for each task using the Fibonacci sequence;
3. The priority of the tasks.

The work is planned in such a way that the team expects to accomplish approximately 35 points per iteration. They can then update the release plan with information for each iteration in release 1, as shown in Figure 10-3.

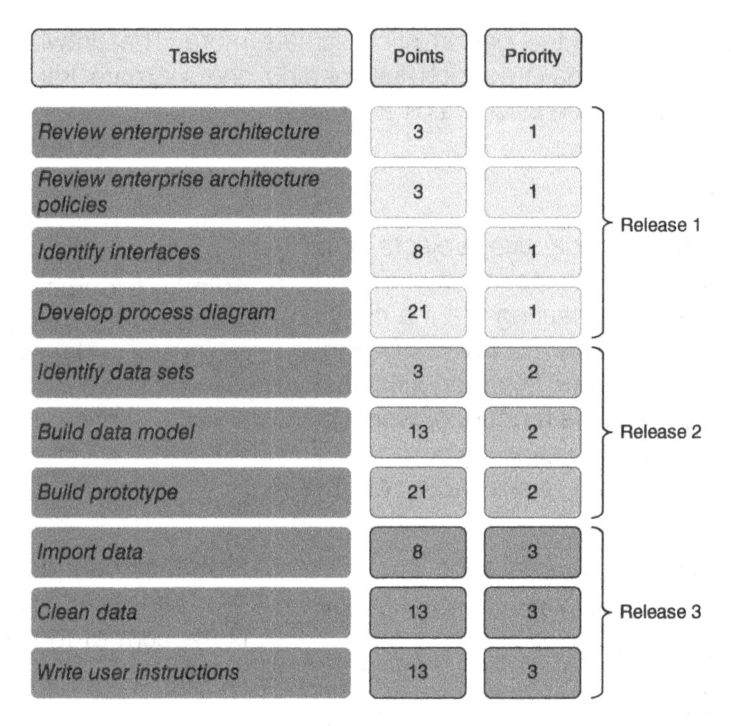

FIGURE 10-2 Prioritized backlog for release 1.

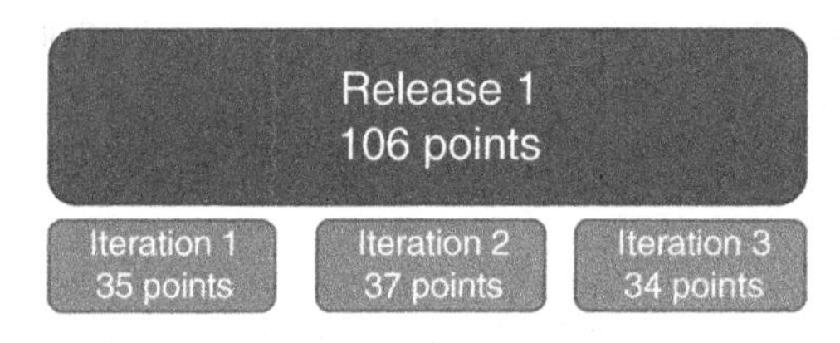

FIGURE 10-3 Updated release 1 plan.

Note: If the team is working with user stories, the points are called "story points." A story point is the unit of value used to estimate the effort needed to complete a user story. Story points are often used with the Fibonacci sequence.

Uses and Benefits

The benefit of this method is that after a team has worked together a while, they get very good and very fast at estimating work. The process is quicker than other estimating methods and is fairly accurate. If the team is using iterations, after a few iterations the team knows how much work they can do in each iteration. If a team using the Fibonacci sequence consistently delivers about 35 story points of work in an iteration and a release has approximately 130 story points, they can release in 4 iterations.

Wideband Delphi

Wideband Delphi is similar to relative estimating, but it uses a group of experts (usually the team members) to develop the estimates. There may be several rounds of estimating before there is a general consensus on the estimate.

Wideband Delphi: Working with a group of experts to develop individual estimates, discuss, and reestimate until consensus is reached.

Note: Wideband Delphi mostly uses the Fibonacci sequence to estimate duration.

Planning poker is a common form of wideband Delphi. In planning poker, each person has a set of cards with each card having one of the Fibonacci sequence numbers on them: 1, 1, 2, 3, 5, 8, 13, and so forth. A facilitator keeps the group focused and moving forward. A product owner gives a brief description of the backlog item, and the team has the opportunity to ask questions to gain clarification about assumptions and risks. Then the facilitator asks the team to pick their estimate from their set of cards. At the same moment everyone flips their card over.

If there is consensus about the work, the team moves on to the next element of work. If not, the outliers (those with significantly higher or lower cards) are given time to discuss why they rated the work that way. Figure 10-4 shows an example of a set of cards with an outlier.

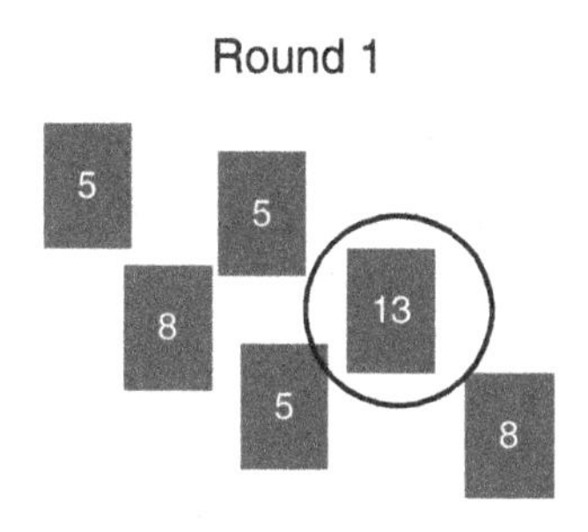

FIGURE 10-4 Wideband Delphi with outlier.

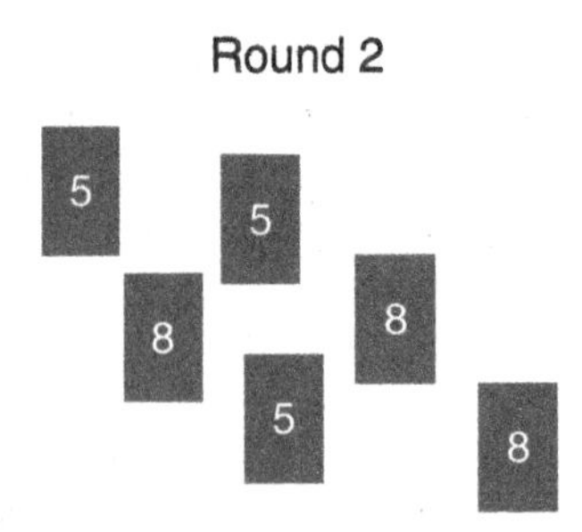

FIGURE 10-5 Wideband Delphi with rough consensus.

After the discussion there is another vote. This continues until all numbers are within one card ranking of each other, as show in Figure 10-5.

If there is a rough consensus, such as three votes of 5 and three of 8, the larger number is chosen.

Bottom-Up Estimating

Bottom-up estimating aggregates the detailed estimates from each activity to determine the overall total. This type of estimating is more common in cost estimating, but the same concept can be used in estimating effort as long as you take the network diagram sequence into consideration.

> **Bottom-up estimating:** Summing estimates from individual work packages to arrive at an overall estimate.

The benefit of this method is that it is usually very accurate. However, it is time consuming because you have to have a detailed understanding of the work. Therefore, you only know the bottom-up estimate after you have fully decomposed scope and are ready to baseline. Hence, it is not used with Agile projects since the scope evolves.

Basis of Estimates

Once you have developed your estimates, it is a good practice to document your basis of estimates. The basis of estimates provides supporting detail about the estimate. It demonstrates transparency about how the estimate was developed and any factors that influenced the estimate. Common elements when documenting the basis of estimates include:

- How the estimate was developed;
- Assumptions and constraints associated with the work and the estimate;
- The range of estimates;
- The confidence level in the estimate;
- Risks that affect the estimate.

On smaller projects this information may be included in the notes. For larger projects you will want to include more robust documentation. It is worthwhile to note that a change to assumptions or constraints for the estimate usually results in a change in the estimate. For example, if you assume a system engineer contractor will be available for four months and will cost $150 per hour, and then you find out there are no local system engineers available and the best option you can find is $180 per hour, you will have a variance. Your original assumption was not accurate, and therefore your estimate will go up.

ESTIMATING THE BUDGET

Of the estimating methods described previously, analogous, parametric, multipoint, and bottom-up are the methods used for cost estimating. Once you have your cost estimates, you can build a project budget. A budget is a time-phased estimate of costs. You can build a budget for a work package, control account, deliverable, phase, or the project as a whole. You can see how to use the cost and duration estimates for the foundation work package for the winery. The foundation work package has these tasks.

Budget: The approved time-phased estimate for project work.

1.2.1.2. Foundation
  1.2.1.2.1. Excavate
  1.2.1.2.2. Grade
  1.2.1.2.3. Install footings
  1.2.1.2.4. Place forms
  1.2.1.2.5. Install rebar
  1.2.1.2.6. Pour concrete

The network diagram with weeks across the top is shown in Figure 10-6.

The concrete contractor gave you the following estimates that he developed using the bottom-up method. He estimated the hours, the hourly rate, the cost of materials, and the cost of using the equipment.

Foundation	**$42,700**
Excavate	$12,000
Grade	$ 8,200
Install footings	$ 2,500
Place forms	$ 3,100
Install rebar	$ 6,700
Pour concrete	$10,500

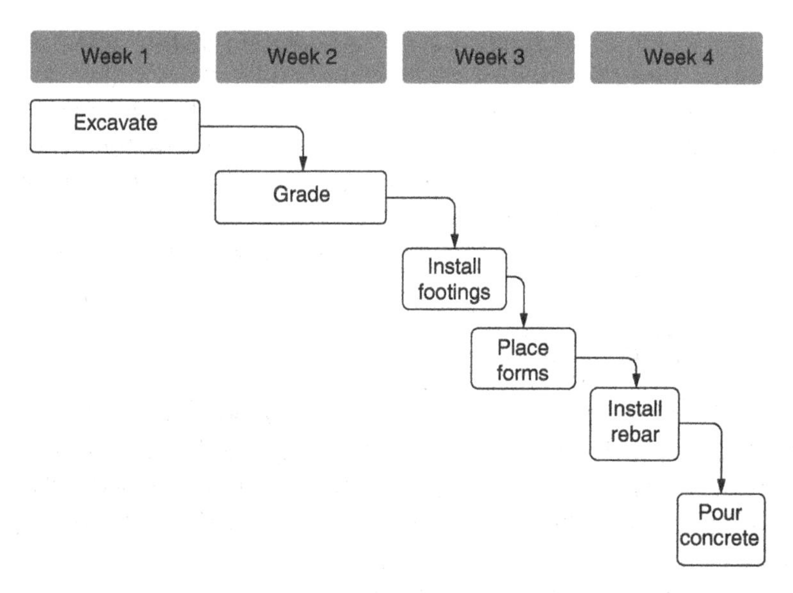

FIGURE 10-6 Foundation work package network diagram.

TABLE 10-4 Budget Worksheet for the Foundation

Foundation	Week 1	Week 2	Week 3	Week 4
Excavate	12,000			
Grade		8,200		
Install footings			2,500	
Place forms			3,100	
Install rebar				6,700
Pour concrete				10,200
Cost per week	12,000	8,200	5,600	16,900
Cumulative cost	12,000	20,200	25,800	42,700

Using the network diagram and the cost estimates you can use a spreadsheet to build a budget for the work package by showing the costs over time. Table 10-4 shows the tasks down the left-hand column and the weeks across the top row.

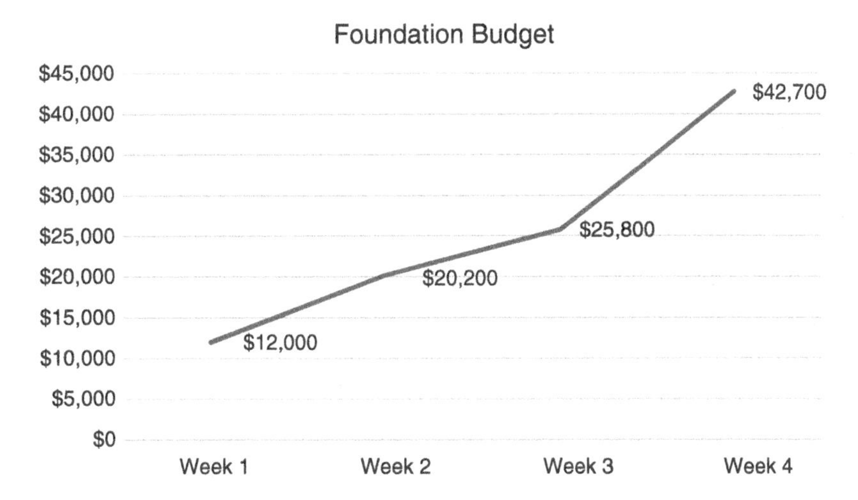

FIGURE 10-7 Budget chart for the foundation.

Notice that the second-to-bottom row shows the cost per week. The bottom row shows the cumulative cost. You derive the cumulative cost by adding the previous week's cumulative cost to the current week's cost per week.

To show the budget as a chart, select the data and insert a line chart, like the one shown in Figure 10-7.

By developing the budget this way, you will be able to analyze your costs for each deliverable and for each time period.

SUMMARY

In this chapter we discussed how the range of an estimate gets narrower as more information is known about the project. We described and demonstrated six ways to estimate effort, duration, resources, and costs. Analogous, parametric, multipoint, and bottom-up are typically used for predictive work. Affinity grouping and wideband Delphi are used for adaptive work.

Using the cost estimates you can build a budget to show the time-phased cost of work. To build a budget you list the work in a column—depending on the level of detail, this may be tasks, work packages, or deliverables. Then arrange them in a network diagram over time. Enter the estimate for each element of work in the corresponding cell. Sum the cost for each time period, and then calculate a cumulative cost on the bottom row. When that is done, you can turn the data into a chart to show the time-phased budget.

Key Terms

affinity grouping
analogous estimating
bottom-up estimating
budget
Fibonacci sequence
multipoint estimating
parametric estimating
point estimate
wideband Delphi

11

Stakeholder Engagement

The technical aspects of project management, such as working with the scope, schedule, and cost, are important, but a key factor in delivering hybrid projects successfully is the way you engage with stakeholders. Successful stakeholder engagement differentiates outstanding project leaders from good project leaders. To be successful in stakeholder engagement, you need to be an excellent communicator. This entails written and verbal communication competencies as well as meeting management skills.

In this chapter we'll identify, analyze, and record information about project stakeholders. Then we'll look at planning for successful engagement using different communication methods. A stakeholder communication plan will record our engagement and communication activities.

IDENTIFYING YOUR STAKEHOLDERS

Stakeholder: People and groups who can affect your project or are affected by your project.

Stakeholder engagement starts at the very beginning of a project and continues throughout the end of the project. As soon as you have your project vision or charter, you should start identifying those people and groups who are affected by your project, have expectations or needs associated with your project, or can impact your project. These people are your project stakeholders.

The project charter and business plans are good places to begin to identify stakeholders. The purpose, project description, and benefits will help you identify people who will be involved with your project or will have a vested interest in the outcomes. If you're outsourcing part of your project or purchasing goods or services, you should look at contracts and procurement documents. For example, if you need to bring in consulting help or perhaps temporary workers, the consultant, temp agency, and the workers are all stakeholders. They can influence the success of your project.

A generic list of stakeholders that are present on most projects includes:

- Customer;
- End users;
- Sponsor;
- Product owner;
- Project manager;
- Scrum master;
- Project team;
- Functional managers;
- Project or program office (PO or PMO);
- Portfolio steering committee.

If you want to look a little deeper, consider

- Regulatory agencies;
- Competitors;
- Internal systems (systems can limit your options or be affected by your decisions).

Stakeholder identification and engagement takes place throughout the project, so once you have identified your initial set of stakeholders, that doesn't mean you are done. You should continue to stay attuned for new stakeholders.

ANALYZING STAKEHOLDERS

Once you identify your stakeholders, you can start to analyze how they will influence or be influenced by your project. There are several variables you can use to assess stakeholders on your project, such as:

- Power or influence: describes the degree of impact a stakeholder has on the project;
- Interest: identifies the level of concern or care a stakeholder has for the project;

- Attitude or support: describes the degree to which a stakeholder is in favor of the project;
- Role: describes how a stakeholder participates with the project;
- Awareness: identifies stakeholder's awareness and support for the project.

You can analyze these variables using 2 x 2 matrixes, cubes, and tables. Several examples are shown below.

Grids and Matrixes

One simple way to consolidate stakeholder information is to create a grid that maps stakeholder positions based on multiple variables.

2 x 2 Matrix

A 2 x 2 matrix analyzes two variables. Fox example, say you want to figure out who has the most influence and who is the most supportive of your project. You can create a power/support grid as the one shown in Figure 11-1.

This grid shows that for those stakeholders with high interest and high power, you should engage actively. Sponsors and customers fall into this quadrant. Stakeholders with high power and low interest, such as senior managers who aren't directly involved with the project, should be kept satisfied. If they request information or have questions, do what you can to meet their needs. Stakeholders with high interest and low power, such as end users, should be kept informed of project progress and other information that would interest them. Those stakeholders with low interest and low power should not consume too much of your time. Monitor them to see if they shift to another quadrant.

2 x 3 Matrix

The 2 x 3 matrix looks at two variables, but goes into a bit more detail than a 2 x 2 matrix. Like the 2 x 2 matrix, it considers each stakeholder's power, but it classifies stakeholders

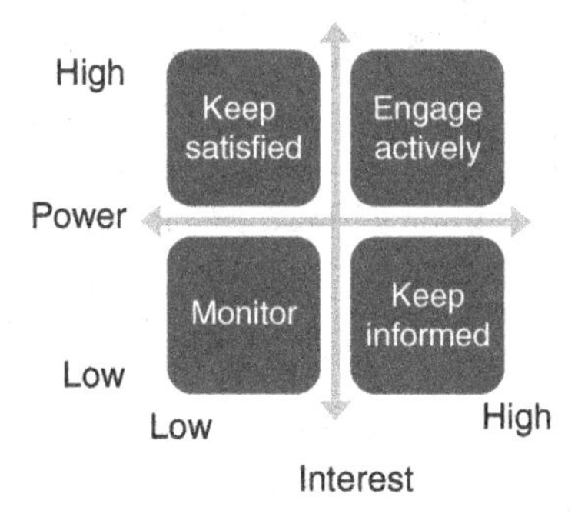

FIGURE 11-1 Power/support grid.

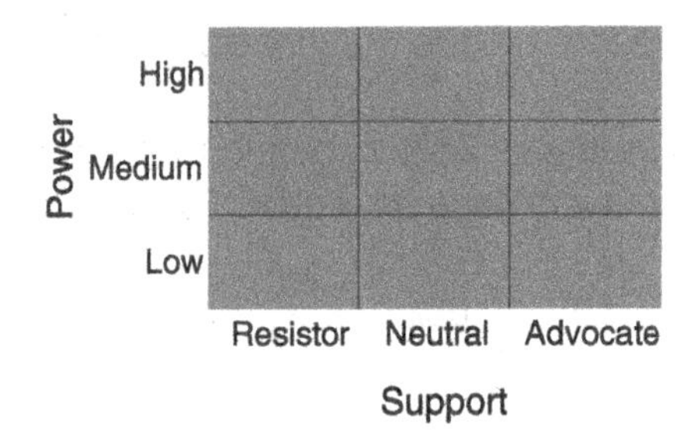

FIGURE 11-2 Power/support 2 x 3 matrix.

as having high, medium, or low power. It also assesses support, classified as resistor, neutral, or advocate.

- A resistor actively and vocally does not support your project. They can cause disruptions to your project and take up your time in trying to work with them. You want to find ways to show the resistor the WIIFM (*What's in it for me?*) in supporting your project. You may also try to find ways to minimize their influence on the project.
- A neutral stakeholder doesn't support or detract from your project. Ideally, you want them to support your project, but they won't do any real harm by being neutral.
- An advocate is actively and vocally in favor of your project. Try to find ways to encourage them to voice their support, and try to maximize their influence.

Figure 11-2 shows a power/support 2 x 3 matrix.

Stakeholder Cube

A stakeholder cube allows you to classify stakeholders according to three variables. The cube depicted in Figure 11-3 shows power, attitude, and interest.

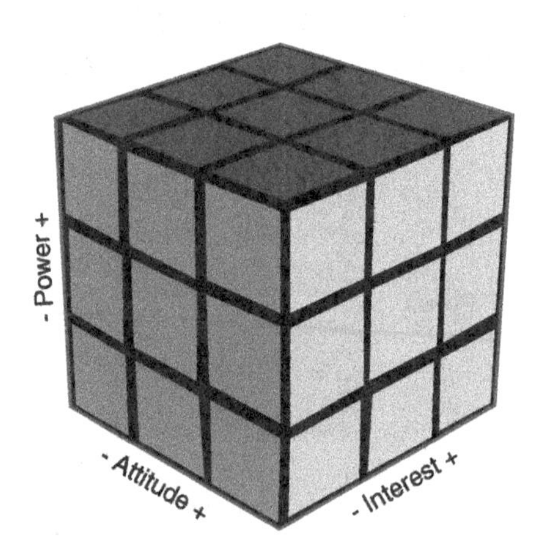

FIGURE 11-3 Stakeholder cube.

You can create any kind of grid, depending on what you think is best for the project, but the overall goal is to use this analysis to develop a strategy for engaging effectively and efficiently with stakeholders.

Analyzing Stakeholders by Role

Another way to categorize stakeholders is by the type of role they have on the project, such as driver, supporter, and observer.

Driver: A driver is person who has the authority to determine the direction of the project and make decisions that affect the project. A project sponsor is an example of a driver.

Supporter: A supporter is a person who provides assistance or resources used by the project to meet objectives. Department managers who are providing team members to work on the project are supporters.

Observer: An observer is a person who does not interact with, or directly influence, the project. End users or other stakeholders who have low influence are observers.

Direction of Influence

The way you engage with stakeholders often depends on where they are on the organizational chart. This affects the direction of their influence. For example:

- **Upward:** Upward includes people such as sponsors, senior management, and the PMO.
- **Sideways:**. Sideways influence refers to peer relationships.
- **Outward:** Outward refers to external stakeholders including regulators, suppliers, and end users.
- **Downward:** Downward refers to the team, though on adaptive and hybrid projects this is not as valid as on a predictive project in a hierarchical organization. In adaptive and hybrid projects you might classify the team as sideways communication.

Awareness and Support

You can classify stakeholders by their level or awareness and support using these categories.

- **Unaware:** The stakeholder is not aware of the project nor its outcomes.
- **Resistant:** The stakeholder is aware of the project but does not support it.
- **Neutral:** The stakeholder is ambivalent about the project and its outcomes.
- **Supportive:** The stakeholder feels favorably about the project and its outcomes.
- **Leading:** The stakeholder is actively engaged in promoting the project.

These examples are a few of many ways you can analyze stakeholders. The needs of your project will determine the best way for you to analyze and categorize stakeholders. If you haven't worked with any of these ways of classifying stakeholders, try a few out and see what works best for your projects.

STAKEHOLDER REGISTER

A stakeholder register is a useful tool to record information about your stakeholders, especially when you have a lot of stakeholders with varied interests. The stakeholder register in Table 11-1 shows an example from the Dionysus Winery project.

Stakeholder register: A document that records relevant information about project stakeholders.

Additional fields for a stakeholder register can include contact information, requirements, and prioritization. Of course, you should tailor the information in your stakeholder register to meet the needs of your project.

You may have stakeholders who come and go from your project depending on the phase, location, or other variables. Therefore, your stakeholder register is a dynamic

TABLE 11-1 Stakeholder Register

Name/role	Expectations	Power	Interest	Attitude
Tessa Barry, Sponsor	Delivery on time, on budget with minimal changes and issues	H	H	Advocate
Tony Dakota, project manager	Competent team, executive support, minimal changes	M	H	Advocate
Sophie Rojas, scrum master	Competent team, project manager, product owner and sponsor support	L	H	Advocate
Angelo Romero, product owner, production manager	Input into all aspects of storage renovation and winery management system	M	H	Advocate
Viniculturist: grape and vine expert	Input into all aspects of grape and vineyard decisions	M	H	Advocate
Enculturist: wine-making expert	Input into all aspects of wine production	M	H	Advocate
Wine club members	Good wine, good value, education, social opportunities	M	M	Advocate
Getaway guests	Fun getaway, good wine, good food, nice facilities, and entertainment	L	M	Neutral
Local residents	Good wine, good food; some concerns about noise and traffic	L	L	Neutral

document that should be updated throughout the project. It is a good practice to review it periodically to see if there are any changes to the power, interest, influence, or expectations of stakeholders.

PLANNING FOR SUCCESSFUL ENGAGEMENT

The way you engage with stakeholders can mean the difference between a successful project and a failed project. The larger your project, the more time you will spend managing stakeholders and their engagement level, and the more critical this process is to the success of your project. Imagine a project with city and county government, the public, the department of transportation, police, and various and sundry other stakeholders. A great deal of your time will be spent engaging with and balancing the various needs and interests of all these very influential stakeholders!

Just as you will be identifying and analyzing stakeholders throughout the project, you will also be engaging and communicating with them throughout the project as well. For each stakeholder or stakeholder group you identify, you will want to think about how best to engage with them. For a project with few stakeholders, you may already know what you need to do. The larger and more complex your stakeholder community gets, the more likely you will need to plan your engagement strategy.

You may want to create a stakeholder engagement plan to organize your strategy. You can start with looking at each stakeholder's current level of awareness and attitude and think about what the desired levels are. Then determine whether you want to take steps to increase a stakeholder's awareness or support. For example, you can develop a strategy based on the 2 x 2 or 2 x 3 grid cells from your stakeholder analysis. You might look at how to move a stakeholder from high power/low support to high power/high support. If you used the stakeholder analysis model of unaware up through leading, you could document strategies and steps to get from neutral to supportive or even leading.

There are four things to remember about developing stakeholder engagement strategies:

- **Be proactive:** Take time upfront to determine how to make the most of your supportive stakeholders and also how to minimize the potential damage from your resistant stakeholders.
- **Be sensible:** You won't be able to shift everyone's attitude or interest. Prioritize your efforts on those stakeholders who can do the most good or the most damage. Some of your strategies may be politically sensitive, so you should use discretion on what you put in writing.
- **Communicate:** The main tool you have to influence stakeholders and manage their expectations is communication. Plan your communications to ensure they have the most impact.
- **Gain support:** Find ways to align stakeholders' expectations with the project. Show stakeholders how it can benefit them to support your project.

You may need to try multiple strategies to work with some of your stakeholders effectively. Stakeholder engagement doesn't end until the project ends, so stay flexible and responsive.

PLANNING PROJECT COMMUNICATION

Communication and stakeholder engagement work together. In fact, it has been said that 90% of your job is communication! Communication occurs throughout a project in all kinds of ways. Much communication is informal, including hallway conversations, cubicle conversations, and networking. These are all examples of informal verbal communication.

Formal verbal communication occurs with presentations or briefings to the steering committee, customer, or sponsor. Your kickoff meeting and team status meetings are additional examples of formal verbal communication.

However, verbal communication isn't the only way we communicate. A lot of communication on a project is written. This can be formal, such as contracts, status reports, defect lists, project plans, and project documents. Written communication can also be informal, such as a brief note to a team member, an inquiry via email, or a follow-up reminder that something is due soon.

Some methods fall into multiple categories. Table 11-2 shows some of the many ways project managers communicate:

The method you choose for communicating will be affected by whether you're communicating vertically through the organization with sponsors or subordinates, or

TABLE 11-2 Formal and Informal Communication

	Formal	Informal
Written	Status reports Project plans Project documents Contracts Intranets Collaboration tools	Memos Emails Intranets Collaboration tools Text messaging Instant messaging
Verbal	Meetings Brainstorming Problem solving Presentations Briefings Demonstrations Videoconferencing Teleconferencing Networking	Conversations Videoconferencing Teleconferencing Networking

horizontally with your peers. You should also consider whether the communication is internal or external and whether it is official (that is, the company's position on an issue) or unofficial (communication not sanctioned or approved by the organization). All these variables affect the means and methods you use to communicate.

Communication Methods

Another way of classifying communication methods is how information is accessed. This includes push, pull, and interactive communication.

- Push communications are sent to stakeholders. This includes memos, emails, reports, voice mail, and so forth.
- Pull communication involves information sought out by someone. A person actively searches for the information. This can include a team member going to an intranet to find templates, internet searches, and accessing online repositories.
- Interactive communication is the exchange of information. Interactive communication occurs in conversation, phone calls, meetings, and the like. This is the most effective form of communication.

Some general guidelines are:

- Interpersonal communication usually occurs through interactive channels.
- Small group communication is usually interactive or push communication.
- Public and mass communications may be push or pull communication.
- Networking, both in person and online, can be formal or informal.

Communication Technology

Another factor in communication is the technology you have available, especially if you have a team with internal team members and external contractors. Also, consider whether people are centrally located or geographically dispersed. Here are some technology factors that will shape your communication.

Urgency: Do people need information immediately, or can they wait? If you are working on a construction project in a hazardous environment, such as an area prone to hurricanes or tornadoes, you need the ability to communicate immediately with the construction site and the workers who could be affected.

Technology availability: Does your organization have the technology and systems in place to support your communication strategy? Will there need to be significant changes or modifications to the current infrastructure? A project that requires some of the staff to work remotely for several days would require staff to have laptops. If they are normally in the office full time and have only desktop computers, consider the matter up front.

Ease of use: You can set up a really cool website to communicate with your team. It might have document storage, messaging capability, version control, and all sorts of neat functionality, but if it's not intuitive to the audience and easy to use, no one is going to use it. When it comes to communication, remember: Keep it simple!

Expected staffing: Check that everyone has the same platform or the same access to information. This is particularly relevant when working with contractors, who might work on a different technology platform. For example, say you hire a marketing firm for part of your project; that firm uses Apple computers, but your firm uses PCs. Staffing your project with members from the marketing firm affects project communication. You also have to determine what information to share with them versus what is proprietary.

Duration: For projects that are three years or longer, you can count on at least some of your communications technology changing. You might not know how it will change, but it would be wise to keep that in mind. With collaboration tools and cloud applications evolving rapidly, predicting how people will communicate in the future is challenging.

Project environment: The environment can be as simple as part of the team being virtual, or it can include more complex factors such as team members located in different time zones, countries, or even extreme geographic locations, such as in the Arctic Circle or the Mojave Desert.

Sensitivity and confidentiality: Some information is business sensitive and should not go outside the team or outside the company. If you have external stakeholders, you will need to determine which information they can be privy to and which they cannot. For sensitive information you may need stakeholders to sign an NDA (non-disclosure agreement).

STAKEHOLDER COMMUNICATION PLAN

With all the communicating going on, it is useful to have a communication plan. A standard communication plan has information such as:

- Message;
- Audience;
- Method; and
- Frequency.

A more elaborate plan could include:

- Sender;
- Definitions and abbreviations;
- Information flowcharts;
- Constraints and assumptions; and
- Templates.

Since communication is for stakeholders, it can be useful to combine the stakeholder register with the communication plan. Doing this keeps all the information about stakeholders and the information they need in one document. Table 11-3 shows a sample stakeholder communication plan for the Dionysus Winery.

TABLE 11-3 Stakeholder Communication Plan

Name/Role	Expectations	Power	Interest	Attitude	Message	Method	Frequency
Tessa Barry, sponsor	Delivery on time, on budget with minimal changes and issues	H	H	Advocate	Status reports Risks, issues	Meeting Meeting	Monthly As needed
Tony Dakota, project manager	Competent team, executive support, minimal changes	M	H	Advocate	Progress reports, Team meetings	Email Meeting	Weekly Weekly or as needed
Sophie Rojas, scrum master	Competent team; project manager, product owner, and sponsor support	L	H	Advocate	Progress meeting, Iteration planning, Retrospective	Stand up Meeting Meeting	Daily Iteration Iteration
Angelo Romero, product owner, production manager	Input into all aspects of storage renovation and winery system	M	H	Advocate	Iteration planning, Demonstration Blockers/ impediments	Meeting Meeting Meeting	Iteration Iteration As needed
Viniculturist, grape and vine expert	Input into all aspects of grape and vineyard decisions	M	H	Advocate	Schedule, Team meetings	Email Meeting	As needed As needed
Enculturist, wine making expert	Input into all aspects of wine production	M	H	Advocate	Schedule Team meetings	Email Meeting	As needed As needed
Wine club members	Good wine, good value, education, social opportunities	M	M	Advocate	Information on wine club	Mail or email	Weekly, then as needed
Getaway guests	Fun getaway; good wine, food, facilities, and entertainment	L	M	Neutral	Marketing information	Email and direct mail	Campaigns as needed
Local residents	Good wine, good food; some concerns about noise and traffic	L	L	Neutral	Marketing information	Email and Direct Mail	Campaigns as needed

SUMMARY

In this chapter we talked about the importance of identifying and assessing project stakeholders. We described several ways to analyze stakeholders using grids and cubes and by evaluating roles, influence, and awareness. This information can be recorded in a stakeholder register.

Communication is the main way we engage with stakeholders. We looked at different methods of communication and the impact that technology has on communication. This information influences how we communicate with stakeholders. A communication plan can document information such as who needs information, the information they need, how it will be delivered and when it will be delivered. A stakeholder communication plan combines the information in the stakeholder register with the communication needs.

Key Terms

stakeholder
stakeholder register

12

Maintaining Stakeholder Engagement

Planning for communication and stakeholder engagement is just the first part of successful stakeholder interaction. Throughout the project you will be managing stakeholder expectations, addressing concerns, and resolving issues. These activities require effective communication skills.

In this chapter you'll learn how to build your communication competencies to effectively engage with stakeholders. This includes giving and receiving feedback and recognizing and addressing communication blockers. Because meetings are such a big part of leading projects, we'll discuss some of the more common meetings for adaptive and predictive projects.

ENGAGING STAKEHOLDERS

The best way to maintain satisfied stakeholders is to communicate effectively and efficiently. Effective communication is providing the right information, in a timely fashion, in a format that works for people. Efficient communication is providing the necessary information to the right people. In other words, send relevant information only to the people who need it. Responding to a message with "reply to all" with a rambling monologue is not effective or efficient. Following your stakeholder communication plan should help you maintain effective and efficient communication practices.

There are a few more guidelines to help maintain good relationships with your stakeholders. The first is to seek stakeholder input when planning the project and incorporate their feedback into the project plan. Stakeholders are much more likely to support decisions and choices when they had a say in the matter.

Next, stay in touch with people. Talk to your stakeholders throughout the project and make sure there is a common understanding of the project scope, benefits, and timeline. Open discussions will help you manage their expectations.

When concerns arise, try and address them before they become an issue. This may mean having conversations with some stakeholders to explain why the scope they wanted is not part of the project, why they can't have what they want in the time frame they want it, or why you require their staff for specific periods of time. Addressing concerns before they become issues is preferred to letting things fester and escalate.

If, despite your best efforts, a stakeholder has a concern that rises to the level of an issue, address it as soon as possible. You may need to escalate the issue to someone with more authority than you or ask them to submit a change request for review, or you may be able to handle the issue yourself. However you choose to address the issue, you're going to need to employ the communication competencies described in the next section.

COMMUNICATION COMPETENCE

As project leaders we are frequently in the spotlight. Therefore, we need to be proficient in all aspects of communication. Some foundational aspects of communication competence are:

- **Know your audience:** A little homework about your audience goes a long way in establishing your credibility and saving yourself from embarrassment.
- **Establish multiple communication channels:** People absorb information in different ways. Some people operate better by listening to information, others by seeing it. Still others absorb best by being actively engaged in activities. Likewise, people are comfortable sharing information or providing feedback in different ways. Make it easy for stakeholders to share information by asking for feedback in meetings and discussions or providing your email and phone number.
- **Use face-to-face communication when appropriate:** You don't need face-to-face communication to relay simple information, such as the date for the next status meeting. When you need to discuss complex concepts, solve problems, or address sensitive information, meeting face-to-face is the best way to communicate. In person is best, but if you can't meet in person, then videoconferencing is the next best option.

- **Be aware of nonverbal communication:** When you are face-to-face, you have access to a lot of information that you don't have over the phone or by email. Pay attention to tone of voice, eye contact, facial expressions, indications of disinterest, discomfort, irritation, and so forth. Oftentimes how people say something is more telling than the words they use.
- **Communicate at the proper time:** Get a sense of whether it is a good time to communicate with someone. Sometimes you can tell this just by body language and facial expression—someone who is scowling and being very abrupt is probably not in the right space to hear what you have to say. Even if their body language seems okay, it never hurts to check in by asking, "Do you have a few minutes to talk?" or "I need to talk with you about something; when is a good time?"
- **Reinforce words with actions:** You can't tell your team, "This is a safe space" and then be rude, condescending, or impatient. Your actions must align with your words and should reinforce or demonstrate what you are trying to communicate.
- **Use simple language:** The best way to make sure you are understood is using simple language. Rather than saying, "Eschew obfuscation," just say, "speak clearly"!
- **Use repetition:** If you have something really important to communicate, say it more than once and say it in more than one way. Use multiple communication channels, and even consider asking for someone to repeat what they heard.

When Someone Is Speaking

Now that we've covered some of the basic communication competencies, let's focus on how to be an effective listener. While some of this information may seem basic, listening is a key skill in stakeholder engagement. After all, how can you know what someone wants if you don't have good listening skills?

- **Give your full attention:** Giving someone your full attention includes stopping what you are doing, making eye contact, and focusing only on the person and what they are saying.
- **Reflect on what the person is saying:** Take time to think about and reflect on what someone is saying—especially if you don't agree or if they are presenting a different viewpoint than your own.
- **Be open-minded:** Don't come into a conversation with your mind made up. Be willing to see things from a different perspective and be creative and flexible. Don't judge the person or their message. You may not agree, but at least be open-minded enough to listen and consider their point of view.
- **Listen without thinking about how you will respond:** Part of giving someone your full attention is not thinking about how you will respond. It may take a while for you to gather your thoughts once they are done speaking—that's okay. They will appreciate the time you took to really listen.

- **Summarize what you heard and check for understanding:** A big part of active listening is asking for clarification and summarizing what you heard. Ask the other person if you understood them correctly. This simple step can help avoid misunderstandings, and it lets them know you are listening and care about what they are saying.

When You Are Speaking

You will be speaking in all different situations—sometimes in front of large groups, sometimes in small meetings, and sometimes one-on-one. Mastering these skills can help you be a more effective speaker.

- **Speak clearly:** First and foremost, speak clearly. Don't mumble, don't speak too fast, and don't slur. Articulate your words and take time to let your message be heard.
- **Make eye contact:** When you are meeting in person, making eye contact is a powerful way to keep your listeners engaged. People are more likely to pay attention when you are looking at them. Making eye contact indicates to your audience that you are confident, you know what you are talking about, and that you want to engage with them.
- **Check for body language:** As mentioned previously, nonverbal cues can communicate more than words. If you see someone who looks confused, check in with them. Find out if you need to communicate your message in a different way. If your listeners look bored, either call for a break or liven things up.
- **Find various ways to explain complex concepts:** For information that is challenging to understand or is confusing, find multiple ways to explain it. Try using a metaphor, an example, or a story to get your point across. Ask for feedback so you can clarify any points that people are struggling with.
- **Be concise; don't ramble:** If you can get your point across in 10 words, don't use 50!
- **Check for clarification:** Especially if you are communicating complex information, but even if you aren't, ask people what they heard. Ask if there is anything they would like you to go over again. Make sure people feel safe asking for clarification.
- **Try to understand your audience's perspective:** Trying to understand your audience's perspective is similar to knowing your audience, but it is also about being empathetic and understanding things from their point of view.

When You Are Writing

Make no mistake, good business writing is a skill. That means you may have to practice it, but it can be learned. As a project manager you will be submitting reports, presentations, emails, memos, and other documents. Some of it will be informal and some formal. You should strive to have your writing be as clear and compelling as your speaking. The 7 Cs of good writing will help you remember these guidelines.

- **Clear:** Know what you want to accomplish with your writing. Start with the goal in mind, and organize your message around that goal. Avoid using words that are vague or that can be misinterpreted.
- **Concise:** In business writing, less is more. Don't elaborate with detail unless it is necessary for understanding. If you need to include relevant detail or backup, consider putting it as an appendix so the reader can reference it as needed.
- **Correct:** The fastest way to reduce the reader's confidence in you is to publish information that is inaccurate. Make sure to fact-check your data and review the final document before you send it out. Correct also refers to grammar. If you are not confident with your mastery of grammar, have someone proofread your work.
- **Compelling:** It can be more challenging to engage your stakeholders with writing than presenting information in person. Therefore, you need to make sure your writing is compelling. Write in the first person where possible, and use action verbs. Your readers will be more engaged with a sentence like this: "The team attacked the problem head on and delivered successfully" than with the passive voice of this sentence: "Problems were overcome to reach the goal."
- **Courteous:** Being courteous in your writing involves keeping the reader in mind. This includes presenting information in a well-organized way and making sure it is easy to absorb.
- **Conversational:** Depending on the context, writing in a conversational tone is easier and more enjoyable to read. Of course, you don't want to write contracts or other legal documents in a conversational tone, but try and find an easy flow for your words.
- **Complete.** Make sure the information you are providing is complete. If you have reference material, include it as an appendix or provide a link if you are sending an electronic document. Don't make your audience work too hard to receive your message.

Feedback

Presenting and receiving feedback can be challenging. Throughout your career you will have opportunities for both. Whether you are giving or receiving feedback, do so openly and non-defensively. This includes making sure your nonverbal cues are nonthreatening and open.

When you are giving feedback keep these three tips in mind:

- **Present the situation you are referencing:** Rather than stating, "You always interrupt," try something such as, "Today at the meeting when I was presenting the 4th quarter results you interrupted me twice."
- **Indicate what you would like to see improved or altered:** If you are looking for a change in behavior, let them know what you like to see. Continuing with the example above, you could ask the person to jot down their questions and wait until you ask for questions or feedback. You could also ask them to raise their hand.
- **Let the person know the reasons you are looking for the outcome:** People are more likely to respond favorably to feedback if you explain why you are asking them to do or not do something. For example, you could state that when they interrupt you, you lose your flow and momentum. To avoid this, you would like to have people hold questions until the end of the presentation.

When you are receiving feedback, make it easy for the person to provide you with feedback. Well-intentioned feedback is usually helpful.

- **Keep an open posture:** Pay attention to the nonverbal cues you are exhibiting. Are your arms crossed? Are you scowling or rolling your eyes? Make an effort to make relaxed eye contact with the person. Use your active listening skills by nodding to indicate you hear what they are saying, ask clarifying questions, and summarize what you heard.
- **Don't interrupt or justify:** Allow the person to finish what they are saying. You may even want to check to see if they are done, in a polite way. At that point you can ask questions, comment, or just thank them for providing you feedback.

You don't have to accept or apply all feedback (except perhaps from your boss). Use your judgment about what to do with it. Even if you choose not to act on the feedback, it is good information to have.

Communication Blockers

Now that we've covered communication skills, let's look at what can inhibit communication. Communication blockers are things people do that prevent effective communication.

Using the wrong media: Make sure you use the appropriate medium for the message. For instance, don't send a corrective action notification via email—do it in person, one-on-one.

Making assumptions about your audience: It is better to check your assumptions rather than think you know your audience. For example, sometimes we make the mistake of assuming that people know what we know. If you leave out what you consider introductory information and only address more specific issues, your audience might not understand your message. Having a lack of common understanding leads to ineffective communication.

A receiver in an area with background activity: Background activity can reduce the recipient's ability to hear your message or focus on your message because they are distracted.

Ignoring cultural differences: Cultural differences occur due to different backgrounds, different generations, and even different organizations. For example, if you have a contractor as part of your team, that person doesn't know your organizational culture, such as work hours, quality of work, accountability, and so forth. Have a conversation up front and address any differences in culture.

Stereotyping: Stereotyping reduces our ability to really hear what someone is saying. For example, if you assume all engineers think in a certain way, you may lose out on opportunities to use their expertise.

Displaying emotions and reactive behavior: Part of being professional is not letting your emotions get in the way of your communication. We will cover this in more detail in Chapter 13: "Leadership in a Hybrid Environment." For now, we'll just say that communicating with a person who is letting their emotions affect their behavior is challenging.

Employing selective listening: Selective listening means only paying attention to the parts of the message a person wants to hear. For example, you're presenting a status report to your sponsor, and you state that for the most part, the project is on track, but there are a few areas of concern with regards to the schedule. If the sponsor only hears the part about how things are on track and ignores or avoids the information about the schedule concerns, they are employing selective listening.

Sending mixed messages: Consider an example of a department manager who advocates work-life balance and supports people's time with their families, but then the same manager expects people to work on the weekend. This is a mixed message.

If you are practicing good verbal, listening, writing, and feedback skills but are still having challenges with your stakeholders, check for communication blockers. They aren't always obvious, but they are often the cause of communication breakdowns and stakeholder issues.

PROJECT MEETINGS

Meetings are an integral part of accomplishing work in projects—though sometimes they feel like a necessary evil. Since much of the project team's work gets done in meetings, you should understand the foundations of holding effective meetings. Whether you are using predictive meetings, adaptive meetings, or a little of both, you can make meetings more engaging for participants by following these guidelines:

1. **Make the purpose of each meeting clear:** You can do this by preparing and distributing an agenda before the meeting. It's also a good idea to ask team members if there is anything they want to include on the agenda.
2. **Ensure appropriate participation:** Make sure the people who need to be at the meeting are there and that only people who need to be present are invited. During the meeting encourage participation in discussions, decision-making, and problem solving.
3. **Share meeting management:** Though sometimes you may feel like superman or superwoman, it's nice to be able to delegate some of the work around meetings. Here are some ideas for delegation:
   - Have a team member start off the meeting with an icebreaker or a brief, fun activity;
   - Ask team members to lead parts of the meeting;
   - Have someone take notes.
4. **Seek feedback:** You can get feedback about the process the team thinks they should use to make a decision, see if they want to continue a discussion or postpone it, check to see if people need a break, and so forth.
5. **Making decisions:** When making decisions, be clear about the intended outcome and the decision criteria. Get input from each person. Determine if the decision will be majority rules, the person with the most expertise makes the final decision, or if the project manager has veto power.
6. **Manage conflict:** Don't let misunderstandings or disagreements escalate into a conflict. You may need to step in and put a hold on a conversation until additional information can be gathered and emotions are in check.
7. **Distribute meetings notes:** At the very least send out a list of follow-up action items. It's better if you can capture and send out the highlights of the meetings, key decisions, and action items. Make sure the notes don't ramble. Remember the 7 Cs of written communication—clear and concise is important when distributing meeting notes.

Adaptive teams and predictive teams have very different sets of meetings. In the following sections you will see a description of common meetings that adaptive and predictive teams use.

Adaptive Meetings

Teams that are adopting a rigorous use of Agile use the term "events" rather than "meetings." Events provide a consistent cadence and structure. Each event has a specific purpose, timing, and intended outcome. Because the purpose is known and stable, the team knows what to expect and can stay focused on the desired outcome. The release planning event was covered in Chapter 9: "Adaptive and Hybrid Scheduling." In the next four sections you will find four events that Agile teams use to plan their work, check in with each other, demonstrate their work, and learn/improve their process.

Daily Stand-Up

A daily stand-up meeting (sometimes known as a daily scrum) is exactly what you would expect: a meeting that occurs at the same time every day. The daily stand-up serves to keep the team focused on the iteration goal, surface any issues, promote collaboration, and provide transparency on the work that is occurring. The term stand-up is used to emphasize that the meeting should be kept short. Not all teams choose to stand up—in fact, I have heard of some meetings where people stay in 'plank pose' during the meeting!

> **Daily stand-up:** A brief meeting held by adaptive teams to review progress from the previous day, describe the intention for the day's actions, and surface impediments.

A stand-up meeting is for the team to discuss their work with each other. While other relevant stakeholders, such as the product owner, may observe, they do not participate. In a traditional stand-up meeting each team member answers the following questions:

1. What have you worked on since our last meeting?
2. What do you plan to accomplish today?
3. Are there any impediments (or blockers) that could impede progress?

While the scrum master does not participate (unless they have project work), they attend to stay apprised of the team's progress and to be aware of any impediments they need to address.

There are a few guidelines for a daily stand-up that helps keep them productive:

- Anyone with a task must attend;
- Only those people with tasks actively participate;
- No side conversations;
- Discuss issues and resolve problems after—not during—the meeting.

The daily stand-up is used instead of a weekly status meeting, which is common in predictive projects.

Iteration Planning

Iteration planning is held at the start of each iteration. This is an opportunity for the team, scrum master, and the product owner (aka customer or business representative) to agree on the work that will be done in the coming iteration. There are two parts to the iteration planning event. The first part of the event entails the product owner presenting the prioritized backlog. The product owner describes the work elements, or in some instances user stories, in the backlog. The team has the opportunity to ask questions and get clarity about the work.

> **Iteration planning:** An event to identify, clarify, and estimate work that will be done in the current iteration.

The product owner is accountable for prioritizing the work, and the team is accountable for determining what work they can accomplish in the iteration. There may be circumstances that prevent the team from working on just the highest priorities on the backlog. For example, there may be a dependency on other work in the backlog, or in hybrid projects, on predictive work in the integrated master schedule. Thus, this event allows the team and the product owner to select the work they can accomplish in the iteration.

Once the work is selected, the team and the product owner agree on acceptance criteria. This can involve establishing the definition of "done" that the team will work toward. By defining "done," the team doesn't over-engineer a solution, and they have objective criteria to measure against. If the work is coding-related, the team will write acceptance tests that they will use to test if a user story is done.

For example, if there is a user story that says,

> *"As a server, I want to know if a bottle is out of stock so I can recommend a different wine,"*

the following are tests that could be used to demonstrate that the user story is complete:

- Searching for a bottle of wine that is not in inventory results in an "out of stock" message;
- Searching for a bottle of wine that is not in inventory triggers a suggestion for a similar wine;
- Searching for a bottle of wine that is not in inventory sends a text alert to the operations manager.

Once the product owner and the team reach agreement on the work for the iteration, the definition of "done," and the acceptance tests, the first part of the meeting is over.

The second part of the meeting is where the team decomposes the user stories into tasks, estimates how long it will take to get each task done, and assigns the work to a team member. This part of the meeting only includes the team members, not the product owner. Once the tasks are estimated and assigned, the team sets up their task board for the sprint and gets to work.

Iteration review: An event to demonstrate the work that was accomplished in the current iteration.

Iteration Review

At an iteration review the team demonstrates the work they have done during the iteration. Attendees at the review include the product owner, scrum master, team, and other interested stakeholders. The product owner will either accept the work or provide feedback on what is missing. Based on the demonstration and the feedback received, the product owner may reprioritize the backlog for the next iteration.

Retrospective

A retrospective is an opportunity for the team to reflect on their performance, approach, and results from the iteration. The purpose is for the team to find ways to inspect, adapt, and improve their ways of working. Retrospectives generally address the following questions:

> **Retrospective:** A workshop to review the work product and processes to find ways to improve outcomes.

- What did we do that worked well?
- What areas need improvement?
- What should we try to improve in the next iteration?

Retrospectives often use a framework to guide the process, such as the four Ls.

The four Ls involves giving each team member several sticky notes for posting onto each of the L areas. Then set up a flip chart, whiteboard, or online whiteboard and dividing it into four sections as follows:

Liked: Team members place sticky notes that describe things they really liked about the iteration.
Learned: Team members place sticky notes that mention things they learned during the iteration.
Lacked: Team members place sticky notes that document things they think could have been done better.
Longed for: Team members place sticky notes that describe things they would have liked to have seen or done in the iteration.

Figure 12-1 shows an example of a four Ls retrospective board.

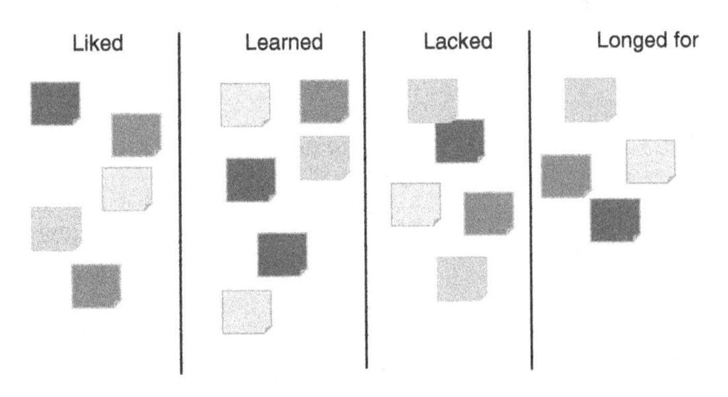

FIGURE 12-1 Four Ls retrospective board.
Source: Cynthia Snyder Dionisio 2017 / John Wiley & Sons.

Another retrospective is a starfish retrospective, as shown in Figure 12-2; it shows sections for "start doing," "stop doing," "do less," "do more," and "keep doing."

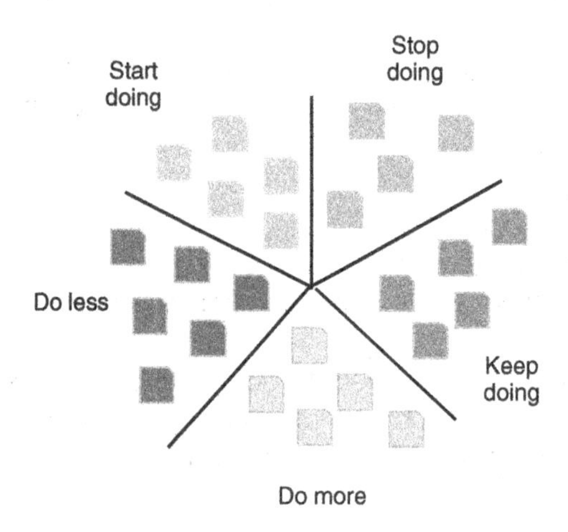

FIGURE 12-2 Starfish retrospective board.

Predictive Meetings

Predictive meetings may have a set cadence, such as weekly status meetings, or they may be held as needed. Having a weekly status meeting just to have the meeting is a waste of time. Therefore, you may only want to have meetings when there are important updates to share or when there is a specific need, such as solving a problem, gaining buy-in, or brainstorming. However, you don't want to go too long in between meetings because people will lose the team spirit and connection to the team. It is a balancing act that varies for each project.

Unlike adaptive meetings, which are primarily self-organized by the team, with predictive meetings the project manager generally manages the agenda, timeline, and participation.

Status Meeting

Status meeting: A meeting to discuss the current progress on the project.

The purpose of a status meeting is to understand the progress that team members have made since the previous meeting. Status meetings are also a good time for the project manager to share news and updates with the team and discuss pending risks and issues. If a topic comes up and it looks like it will require a lengthy discussion, you may want to set up a "parking lot" to handle such topics at the end of the meeting if there is time or address them at a later date.

Lessons Learned Meeting

A lessons learned meeting is held at the end of a phase, at the end of the project, or when a significant change or risk occurs. The purpose of this meeting is to identify actions or approaches that worked well so they can be continued and shared with other teams and to identify actions that didn't work so well so the team can learn from them.

Lessons learned meeting: A meeting to identify and document areas that are working well on the project and areas that can be improved.

Typically, lessons learned are documented in a register or repository so they are available for future reference. Lessons can be categorized by topic, such as stakeholder engagement and risk management, by project phase, or other categories as appropriate.

Lessons learned are different than adaptive retrospectives in that they are not held at a regular cadence, and they are not typically focused on how the team can improve their processes. However, with hybrid projects, you can mix and match and do what works best for your project. Some options include:

- Adopt a retrospective structure, such as the four Ls for a lessons learned meeting;
- Include a question about how the team can improve how they work together at the end of each status meeting.

SUMMARY

In this chapter we discussed the role that communication plays in stakeholder engagement. We started with identifying several aspects of communication competence. We expanded the conversation to look at techniques you can use when others are speaking, when you are speaking, and when you are writing. We also discussed some tips for giving and receiving feedback and some of the main sources of communication blockage.

Because meetings are such a significant part of engaging stakeholders and managing projects, we described overall meeting guidelines and then looked at events used in adaptive projects and meetings used in predictive projects.

Key Terms

daily stand-up
iteration planning
iteration review
lessons learned meeting
retrospective
status meeting

13

Leadership in a Hybrid Environment

Leadership is a critical part of project success. It involves strategic skills, such as communicating the project vision, goals, and objectives, and people skills, such as emotional intelligence, motivation, and team development. Project leaders are responsible for inspiring their teams to reach the intended project outcomes. On a hybrid project, there may be some team members who require a hands-on approach and other team members who are self-organizing. This requires the project manager to be skilled in situational leadership in order to apply the right style of leadership at the right time.

In this chapter we will discuss emotional intelligence as a foundational skill needed to know how to best lead your team. Emotional intelligence includes motivation; however, because this is such a significant topic, we will look at several factors that influence motivation.

On hybrid projects you are likely to have Agile teams that practice servant leadership and are self-organizing. We will look at how you can utilize situational leadership to know how and when to support these practices on hybrid projects. Regardless of whether you have a predictive, adaptive, or hybrid project, there are some key practices involved with developing high-performing teams. These will be covered so you can apply them regardless of the structure of your teams.

EMOTIONAL INTELLIGENCE

According to Dr. Daniel Goleman, "A person can have the best training in the world, an incisive, analytical mind, and an endless supply of smart ideas, but he still won't make a great leader." Emotional intelligence is a defining factor in the difference between leading based on position versus being a great leader.

Emotional intelligence: Awareness of one's own emotions and moods and those of others, especially when leading people.

There are five aspects of emotional intelligence:

- Self-awareness;
- Self-regulation;
- Social awareness;
- Social skill;
- Motivation.

Self-Awareness

Self-awareness is recognizing and understanding your moods and emotions and how they affect you, your job performance, and your team members. Part of self-awareness is being aware of how you are feeling in the moment and not judging those feelings. In this way, self-awareness is similar to mindfulness. Another aspect of self-awareness is identifying what your triggers are—what leads to you getting upset? Depressed? Apathetic? Being able to recognize and understand what you are feeling and why you are feeling that way allows you to regulate how you behave.

Self-Regulation

Self-regulation is the ability to think before acting. It is suspending reactive behavior. When you identify emotional triggers, you can take steps to avoid those situations, but eventually you will find yourself in a situation where you are going to be upset. Therefore, once you identify your triggers, it can be helpful to think about healthy ways to disengage your emotions. For example:

- Take a walk;
- Take a shower;
- Take 10 deep breaths;
- Write down what is bothering you;
- Take a break;
- Take a nap.

Self-regulation can help you make decisions and take actions based on thoughtful consideration rather than emotional reactions. Emotional outbursts reduce safety and trust, whereas practicing self-regulation improves trust on your team.

Social Awareness

Social awareness is being aware of other's viewpoints and emotions. It includes understanding nonverbal cues and displaying empathy. For example, it's pretty obvious if you come across someone who has their arms crossed, is looking down, and not meeting your eyes that they don't want to interact with you. They may be angry, or upset or sad, but their body language is definitely saying, "leave me alone." There are more subtle cues as well, a sigh, mumbling, or speaking slowly without much intonation can also communicate a person's mood. So can smiling, nodding, and gesturing.

As leaders we should be aware of these cues. We may even want to check in with our team members and ask if there is something they want to talk about or something they need help with. Pay attention to their tone of voice, posture, facial expressions, and gestures.

Social Skills

Social skills are about leading and managing groups of people, such as project teams. They also include building social networks and establishing rapport. In this context, establishing rapport means building a harmonious and sympathetic connection with others. As project leaders, we rely on our ability to create rapport with team members, functional managers, senior management, and other stakeholders.

Motivation

There are two types of motivation—intrinsic and extrinsic. Motivation, when considered as a component of emotional intelligence, refers to intrinsic motivation. Intrinsic motivation comes from inside a person or is connected with a task itself. It is based on finding pleasure or meaning in the work instead of the reward.

Intrinsic motivators include achievement, taking on a challenge, or being self-directed. Other examples include believing in the work you do or knowing your work makes a difference. Most projects require creative thought, problem solving, inventing new technologies, and innovating. These types of tasks are more motivated by intrinsic factors, and the people who work in these types of jobs generally enjoy the work they do.

Extrinsic motivation is doing something based on an external reward, such as job title, bonuses, and cash rewards. Winning and avoiding failure or embarrassment are also extrinsic motivators. Extrinsic motivation works well only for a very narrow focus when there are a simple set of rules—for example, performing repetitive work that doesn't require much thought or creativity. Extrinsic motivation is, for the most part, far less effective than intrinsic motivation.

Tip: There are many emotional intelligence self-assessments available on the Internet. Many of them are free; your only investment is time.

MOTIVATORS

Motivation is far bigger than just understanding the emotional intelligence aspect. There are many motivation models, and most of them overlap or describe similar concepts. Most of the intrinsic motivators fit in one of the following four categories.

Autonomy: Autonomy is the ability to direct our own lives. This includes making choices about what we work on and how we spend our time. Autonomy allows us to work on things that are important to us and align with our values.

Competence: Competence is associated with being able to succeed and excel at what we do. It includes mastery, which is about getting better at what we do, especially when what we do matters to us.

Relatedness: Relatedness is connecting with others. Connecting with others increases our sense of happiness and well-being. Connection can include networking, having friends at work, coaching and mentoring relationships, and other forms of social interactions.

Purpose: Purpose is the opportunity to work in service of something larger than ourselves. It includes the feeling that what we do makes a positive difference and that we're contributing.

Motivating Your Team

Because people are motivated by different things, to be effective at motivating your team you need to take time to get to know each person individually and understand what's important to them. You can do this by having candid conversations with your team members. That way you can align their work and incentives with what's important to them. For example, you may have one team member whose autonomy motivation expresses itself by wanting a more flexible schedule, or a schedule that allows them to work from home full or part time. Another team member may see autonomy as being able to sequence their work the way they want.

You can tailor motivation techniques based on a team member's dominant motivating factor. Each of us to one degree or another is motivated by the four motivators described above; however, the relative importance of each motivator varies from person to person.

Table 13-1 shows a few examples of how to tailor the motivation techniques based on a team member's dominant motivator.

TABLE 13-1 Sample Motivation Techniques

Autonomy	Provide team members with freedom and flexibility in their work environment.
Competence	Set stretch goals for team members to achieve. Make opportunities available for professional growth.
Relatedness	Give people an opportunity to be part of a team, and establish opportunities for socializing.
Purpose	Emphasize the difference a person makes in doing their work—make it apparent.

Example of Motivation in the Workplace

Google has long encouraged its employees to devote 20% of their time to side projects they're passionate about. This represents both autonomy and purpose. If people are working with their peers, it also encompasses relatedness. As people work in these areas, they naturally improve, which includes competence.

Many of Google's most successful products have come out of this 20% time. This is one reason Google remains one of the most innovative companies in the world.

AGILE LEADERSHIP PRACTICES

There are two leadership practices that are specific to Agile project teams, servant leadership and self-organizing teams. These practices work well when the project environment understands and appreciates Agile practices. When working in a predictive or top-down management environment they will need to be tempered to suit the situation.

Servant Leadership

Servant leadership: A leadership style that focuses on the needs of the team.

A servant leader focuses on making sure the team has everything they need to get their work done, rather than providing direction. The intended outcome is that team members will feel fulfilled, appreciated, and supported. For teams that use scrum practices, the scrum master is a servant leader.

Some of the ways servant leaders support the team include:

- Ensuring they have everything they need to get their work done. This can include space, equipment, resources, and even decisions.
- Shielding the team from interruptions. One of the challenges in accomplishing work is constant interruptions from managers asking questions, stakeholders wanting to

know the status of work, customers wondering if they can get another feature, and so on. A servant leader is the single point of contact for the team. They protect the team from interruptions and manage all interactions with external stakeholders.
- Removing impediments. An impediment is anything that gets in the way of accomplishing work. Impediments are also called blockers or barriers. Impediments often surface in the daily stand-up meeting. The servant leader tracks down the impediment and works to resolve it.
- Communicating the project vision. The servant leader reminds the team and other stakeholders about the project vision. This keeps people focused on the work at hand. It can keep the team from over-engineering a product, and it can keep non-value adding work (such as excessive documentation) out of the work queue.
- Supporting team members. Servant leaders support team members by providing training, encouragement, motivation, and celebrations.

An environment of servant leadership results in a more efficient and productive workspace and higher-quality work products.

Self-Managing Teams

Self-managing teams pair well with servant leadership. While the servant leader is supporting the team, the team is accountable for delivering the work. The servant leaders may function as a facilitator or coach for the team, but ultimately it is up to the team to figure out how they are going to accomplish the work, rather than a manager deciding for them.

Self-managing teams: Teams that share accountability for the delivery of value.

At the start of the project the team will assess what they need as far as space, equipment, communication, and so forth. They will also determine how they want to work together, such as:

- Communication norms;
- Making decisions;
- Resolving conflicts;
- Addressing technical issues;
- Coming to consensus; and
- Improving their process.

These agreements may be posted in the work environment or may be recorded in a team charter that is signed by each member of the team. Figure 13-1 shows a version of a team charter.

Team charter: A document developed by the team that identifies team agreements and ways of working.

TEAM CHARTER

Project Title: ______________________ **Date Prepared:** ______________________

Team Values and Principles:

1.
2.
3.
4.
5.

Meeting Guidelines:

1.
2.
3.
4.
5.

Communication Guidelines:

1.
2.
3.
4.
5.

Decision-Making Process:

Page 1 of 2

TEAM CHARTER

Conflict Resolution Process:

Other Agreements:

Signature:	Date:
________________	________
________________	________
________________	________
________________	________
________________	________
________________	________
________________	________
________________	________
________________	________

Page 2 of 2

FIGURE 13-1 Team charter.

During the project the team is empowered to organize and estimate their work. This occurs during iteration planning meetings. Then during the iteration, they have the autonomy to accomplish the work, resolve technical challenges, and employ the agreements they documented in the team charter.

Trusting team members to establish their own ways of working increases commitment and motivation. People work harder and take more pride in their work when they are respected and trusted as experts.

Self-managing teams don't always start out smoothly. It can be a goal that takes time to achieve. The servant leader may need to provide coaching at the start and guidance along the way. However, as team members gain an understanding of what needs to be done and how to do it, the leader can step back.

Tailoring for a Hybrid Environment

When choosing a leadership style for a hybrid project, you will need to consider the organizational culture. If the organization embraces a top-down management style, you should follow that approach. You can be open to suggestions, share power, and support the team, but ultimately you will maintain the final decision-making authority.

Conversely, in a very flat organization or an organization that's very collaborative, you may be able to adopt a servant leadership style. There isn't a formula that tells you how to flex your style based on the environment. You'll need to assess the environment, the situation, and the organizational culture. That being said, here are a few guidelines for balancing predictive and Agile leadership styles.

1. Where possible, let team members figure out how to do their own work.
2. Be transparent. When you make a decision, disclose your rationale and reasoning process.
3. Lean toward collaboration and away from autocracy.
4. Share leadership where possible.
5. Use your position as a leader to serve your team.

DEVELOPING A HIGH-PERFORMING TEAM

You've probably had the opportunity to work on a team where everything clicked, even amid turmoil the team worked well together and there was a sense of synergy. You've also probably worked on teams where it was an effort to get anything done—egos were involved, people weren't committed, and it was a chore to get anything done. It's likely that in the first example you were working on a high-performing team. High-performing teams have several traits in common, as you'll see below.

Traits of High-Performing Teams

The first thing that high-performing teams have is a shared sense of purpose. Everyone on the team knows what the intended outcomes are, and everyone is committed to those outcomes. Along with the shared sense of purpose, team members display mutual accountability and trust for each other. Mutual accountability means that everyone on the team takes accountability for the project's success. There's no sense of "I did my job; why didn't you do yours?" The team works together and helps each other out as needed.

Mutual accountability leads to a sense of trust. Team members trust each other not only to get their work done but to consider the greater good of the team, to keep their promises, and to help each other out if needed. High-performing teams collaborate. They don't work in silos. They work through problems and make decisions together, and they do this with open and respectful communication. These characteristics work together nicely. The sense of mutual accountability and trust leads to working together respectfully and collaboratively and vice versa.

High-performing teams share recognition and empower each other. They share in successes and acknowledge each other's contribution to the success.

You've probably worked on projects where something completely unexpected happened and negatively affected the project. Maybe it caused rework, or perhaps a risk materialized, or maybe there was a big schedule slip. These setbacks are hard to come back from. But high-performing teams demonstrate resilience to handle these setbacks, and they demonstrate adaptability to adjust to new ways of working to overcome negative situations.

Finally, these teams have high energy and an attitude of achievement. Team members have a positive outlook—they know that even when times are challenging, together the team can work through the challenges and deliver the project.

Building Relationships

If you want to build a high-performing team, start with investing time in building good relationships. Here are five key elements to developing good relationships with your team.

1. **Establish rapport:** Rapport is a harmonious relationship with people. It doesn't happen in an instant, but you can take steps to foster it. For example, find commonalities with your team members, such as sports you enjoy, favorite foods, or pets. Another way to build rapport is matching tone of voice and choice of words. For example, if you're working with someone who speaks slowly and deliberately, match their speed, don't talk a mile a minute. On a subconscious level people feel more comfortable with those who are similar to themselves.
2. **Shared experiences:** As you go through the project, you'll create shared experiences, which increases rapport. Also be empathetic with what team members are experiencing, especially professionally but also personally, as appropriate.

3. **Value the team and each team member:** You build good relationships when you let team members know that you value them and their contribution. Team members may see things from different perspectives, such as a business analyst and a marketing specialist. Though they see problems from different perspectives, both perspectives have value. By looking at a range of viewpoints that working in a team affords, we can often develop better outcomes for issues that arise.
4. **Be open and transparent:** Being transparent with your thinking and how you make decisions builds trust. As a project manager there will be times when making tough decisions is part of your job. People may not always like your decisions, but if you can be transparent as to why you made a certain decision, it helps people understand. They'll be more likely to trust you if you're open about your reasoning and motivations.
5. **Appreciation:** Finally, it's important to let your team know that you appreciate them. This can be as simple as saying thank you or writing a quick note letting them know you appreciate something they did. It can also include writing a memo to their boss describing the team member's contribution.

Making the investment in employing these elements makes the work you will do with your team more enjoyable. It also sets the stage for them to develop into a well-functioning and high-performing team.

SUMMARY

In this chapter we reviewed some of the interpersonal aspects of working on hybrid projects. We began by looking at the five aspects of emotional intelligence: self-awareness, self-regulation, social awareness, social skills, and motivation. We then explored four intrinsic motivators: autonomy, competence, relatedness, and purpose.

There are two leadership practices that are specific to Agile projects, servant leaders and self-managing teams. We discussed how these practices can be tailored for a hybrid project environment. We finished this chapter by describing the traits of high-performing teams and the important of building relationships with your team to set the stage for a productive team experience.

Key Terms

emotional intelligence
self-managing teams
servant leadership
team charter

14

Planning for Risk

Project risk is a challenging subject because you are dealing with future unknown events. Because projects are unique and temporary, they are by default more risky than sustained operations. Risk, in and of itself, is not bad—it is just uncertainty. However, failure to plan for uncertainty, and failure to take steps to reduce threats is a significant contributor to project failure.

In this chapter we will introduce key concepts in risk management. We'll discuss event-based risks, such as threats, opportunities, and impediments. Then we'll look at risk tolerance and risk thresholds.

For large projects you will need to spend some time considering how you will identify, analyze, and respond to risks. That information is recorded in a risk management plan. This chapter will describe the elements in a risk management plan and show an example from the Dionysus Winery that incorporates those elements.

INTRODUCTION TO RISK MANAGEMENT

Risk: An uncertain event or condition that can have an impact on a project.

Risk management is not a one-time event; it begins at the start of the project and carries through until the project is complete. For projects where there is not a lot of uncertainty and not a lot at stake, you won't need to spend a great deal of time and effort on risk management. However, for projects that are significantly different from previous projects, and when the stakes are high, you will want to invest significantly in risk management activities.

Threat: A risk that will have a negative impact on a project.

Opportunity: A risk that will have a positive impact on a project.

When we speak about risks, we are referring to uncertain events. Often risks are equated with threats; however, our definition of risk includes both negative and positive outcomes. A risk with a positive outcome is called an opportunity.

An example of a risk that is a threat for the Dionysus Winery is a supply chain interruption that could delay the delivery of a de-stemmer crusher for the wine production facility. An example of a risk that is an opportunity is acquiring used wine barrels from a winery that is going out of business.

Issue: A current condition that may have an impact a project.

Sometimes people equate risks to issues. They are similar but slightly different. A risk may or may not occur, whereas an issue is a current situation. If an identified risk does occur, it will affect the project. An issue may or may not affect the project, depending on the issue and how quickly it can be resolved. An example of an issue is the roof tiles for the hotel were discontinued. This is a current situation. If the contractor can find similar roof tiles at a comparable price, it will not affect the project; if he cannot, then it will affect the cost, schedule, or both.

Known risks: Risks you can anticipate and plan for.

Unknown risks: Risks you can't anticipate or plan for.

Another aspect of risks is whether you can reasonably anticipate the event or condition. The risk of the delayed de-stemmer crusher could be anticipated knowing that piece of equipment is manufactured in Italy, and the project is occurring during a time when there are global supply chain issues. That is referred to as a known risk. An example of an unknown risk that could not reasonably be anticipated is that a neighboring winery developed Pierce's disease (a disease where insects transmit a bacteria that kills vines). The disease spread to your vines before it was identified at the neighboring winery.

Impediment: Anything that gets in the way of the team achieving its objectives.

Teams that use an Agile methodology check in with team members every day in a stand-up meeting to see whether there are any impediments or blockers that are preventing them from accomplishing their work. Used in an Agile context, an impediment can refer to a threat or an issue. It can be a known or unknown situation.

RISK TOLERANCE AND THRESHOLDS

Different people and different organizations have varying degrees of comfort in accepting the uncertainty that comes with risk. This is called risk tolerance or risk appetite. For example, you may have one team member who believes that ultimately the team will be able to work together to overcome any obstacle. This person has a high risk tolerance. Another team member may be more risk averse and may prefer to plan everything out and have backup plans in case the original plan fails.

Risk tolerance: The degree of uncertainty an entity is willing to accept.

Risk threshold: The amount of risk exposure an entity is willing to accept before actively addressing a risk.

Neither viewpoint is necessarily good or bad, though the different viewpoints can lead to stress and disagreements. One way to address the differences is to identify risk thresholds for cost, schedule, and performance variances at the start of the project. A risk threshold indicates the point at which the team will take action. For example, a deliverable with a 5% budget variance might indicate the need to identify what is causing the variance and to pay close attention to the spending on that deliverable. A 10% or greater variance may indicate the need to take action to bring the costs back in line with the budget.

Thresholds are good indicators of an organization's comfort level with uncertainty—the lower the threshold, the more risk averse the organization is. If the organization has not established thresholds, it is a good idea for the team to do so at the start of the project. This alleviates disagreements on whether or not the team should develop an active risk response strategy, or whether they should accept the risk.

RISK MANAGEMENT PLAN

The risk management plan is a subsidiary plan of the overall project management plan. It describes how risk identification, analysis, and response planning will be conducted. Information in the risk management plan varies depending on the size, complexity, and importance of the project. The robustness and frequency of the risk management process should mirror the complexity and criticality of the project.

Risk management plan: A subsidiary plan of the project management plan that describes how risks will be identified, analyzed, and addressed.

Elements in a Risk Management Plan

Like everything in project management, tailor the contents of the risk management plan to meet the needs of the project. The following elements are a starting place for a risk management plan. As needed, you can elaborate the contents of the plan.

Roles and responsibilities: Describe the role that various stakeholders play in managing risk. This can include a risk manager for very large projects or the responsibility that each team member has with regard to risk management for small- and medium-sized projects.

Funding: Estimate the funds needed for risk identification, analysis, and response. Define the approach for allocating, using, and recording contingency funds.

Timing: Identify the risk management activities that need to be added to the schedule and determine how often they will occur. Define the approach for allocating, using, and recording contingency for the project schedule.

Risk categories: Identify the major categories of risk on the project and decompose them into subcategories. You can use frameworks or prompt lists for categorizing risk, such as:

- PESTLE (political, economic, social, technological, legal, and environmental);
- TECOP (technological, economic, commercial, operational, and political); or
- VUCA (volatility, uncertainty, complexity, ambiguity).

Definitions of probability: To analyze risks effectively, use a common method of rating the probability of an occurrence. This alleviates misunderstandings or disagreements about risk thresholds and how to analyze risks. You can use a cardinal (numeric) scale or an ordinal (descriptive) scale. A cardinal scale might rank probability as:

- 1–20%
- 21–40%
- 41–60%
- 61–80%
- 81–99%

An ordinal scale could rank probability as:

- Very low;
- Low;
- Medium;
- High; or
- Very high.

Definitions of impact: Similar to establishing a scale for analyzing probability, you can also establish a common method of rating the impact if an event occurs. You can use the same approach for rating impacts as very high, high, medium, and so on—or you can use a cardinal scale from 1 to 5. For large projects it is helpful to have separate definitions of impact for different project objectives, such as schedule, cost, and performance. An example for cost impacts is below.

- Very low: less than 2% cost variance;
- Low: 2–5% cost variance;
- Medium: 5–10% cost variance;
- High: 10–20% cost variance;
- Very high: greater than 20% cost variance.

You would establish similar impact definitions for schedule and performance.

Probability and impact matrix: A probability and impact matrix (P x I matrix) is a grid for mapping the probability of occurrence and the impact on the project if a risk occurs. It is used to rate risks as high, medium, or low. A P x I matrix usually has color gradations of red (high), yellow (medium), and green (low). Figure 14-1 shows a P x I matrix with different shades of gray, where darker gray is high, medium gray is medium, and light gray is low.

This matrix is balanced. The low, medium, and high rankings are relatively equal, as opposed to having more squares indicating high risk as shown in Figure 14-2.

The P x I matrix in Figure 14-2 shows a more risk-averse organization. A P x I matrix with more light colors and less dark would indicate an organization that is more risk tolerant.

The P x I matrixes shown are called a "5 x 5" matrix (5 by 5) because they rate risks in five degrees of probability and five degrees of impact. You can use 3 x 3, 10 x 10 or other

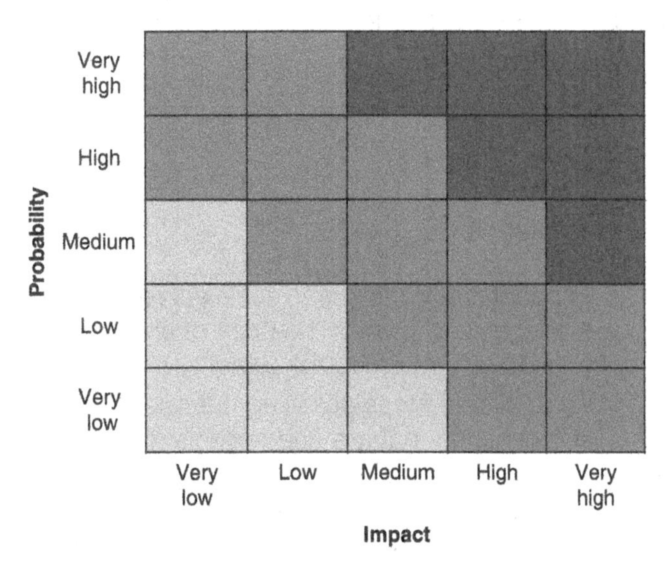

FIGURE 14-1 Sample probability and impact matrix.

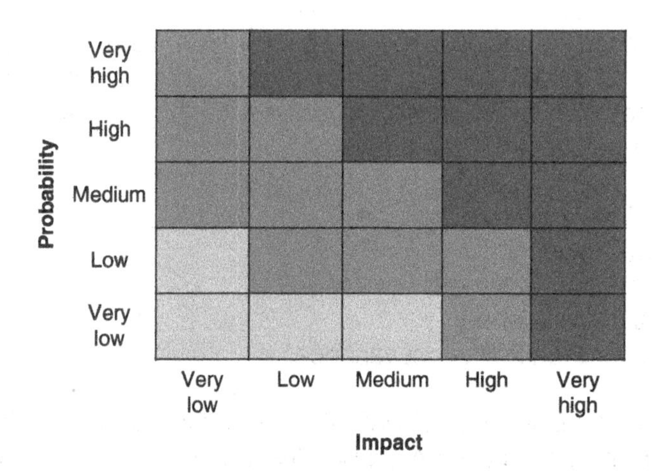

FIGURE 14-2 Risk averse probability and impact matrix.

configurations as well. If you differentiate impacts for each objective, you will have separate P x I matrixes for each objective, as stakeholders may be more sensitive to variations for some objectives than for others. For example, if the schedule is the most important aspect of a project, the P x I matrix for schedule variances may be more risk averse than the P x I matrix for cost.

Sample Risk Management Plan

Now that we have described the content in a risk management plan, we'll take a look at how the Dionysus Winery plan might look. You can see a sample risk management plan for the Dionysus Winery project below.

RISK MANAGEMENT PLAN

Roles and Responsibilities

Tessa Barry	Accountable for ensuring sufficient resources to effectively manage risk. Tessa is the point of escalation for risks that are outside of Tony Dakota's authority.
Tony Dakota	Accountable for establishing the risk management process and ensuring that it is used effectively and consistently.
Angelo Romero	Accountable for identifying and analyzing all risks associated with the wine production and storage facilities and working with the team to respond accordingly.

Viniculturist	Accountable for identifying and analyzing all risks associated with the vineyards and the grapes.
Enculturist	Accountable for providing input into risks associated with the wine production and storage facilities to the degree that they will affect the wine production process.
General contractor	Accountable for identifying and analyzing all risks associated with construction and renovation of all facilities. The general contractor will work with and be the spokesperson for all trades working on the facilities. The general contractor is also accountable for identifying and analyzing risks associated with building permits.
Sophie Rojas	Accountable for identifying impediments for the winery management system during daily stand-up meetings and communicating them to Angelo. Sophie will be the spokesperson for the team.
Human resources	Accountable for identifying and analyzing all risks associated with staffing, training, and the wine club and working with the team to respond accordingly.
Event planner	Accountable for identifying and analyzing all risks associated with the grand opening and working with the team to respond accordingly.
Team members	Accountable for identifying risks in their areas and responding as requested.

Funding

Tessa Barry has allocated 10% of the budget for contingency funds. Tony Dakota has the authority to allocate contingency funds as needed. If needed she has management reserve at her discretion.

Timing

Risk assessment meetings will be held every two weeks. If needed, they will be held weekly before a major deliverable or event.

Risk Categories

The following categories of risk will be used as prompts for identifying risks.

- Construction;
- Regulatory;
- Supply chain;
- Staffing.

Definitions of Probability

- 0–10% = Very low;
- 11–25% = Low;
- 26–50% = Medium;
- 51–70% = High;
- 71–99 = Very high.

Definitions of Impact

	Cost	Schedule	Performance
Very low	Less than 2% cost variance	Up to 50% float used on noncritical path	Barely detectable degradation in a supporting feature
Low	5% cost variance	50% or more float used on noncritical path	Moderate degradation in a supporting feature
Medium	5–10% cost variance	All float used; no impact to critical path	Supporting feature not available
High	10–20% cost variance	Up to 2-week slip on critical path	One key deliverable or feature is nonfunctional
Very high	Greater than 20% cost variance	More than 2-week slip on critical path	More than one deliverable or feature is nonfunctional

Probability and Impact Matrix

A balanced 5 x 5 matrix, will be used for each of the project objectives (cost, schedule, and performance).

This risk management plan will guide and support the team in planning and managing risks effectively throughout the project.

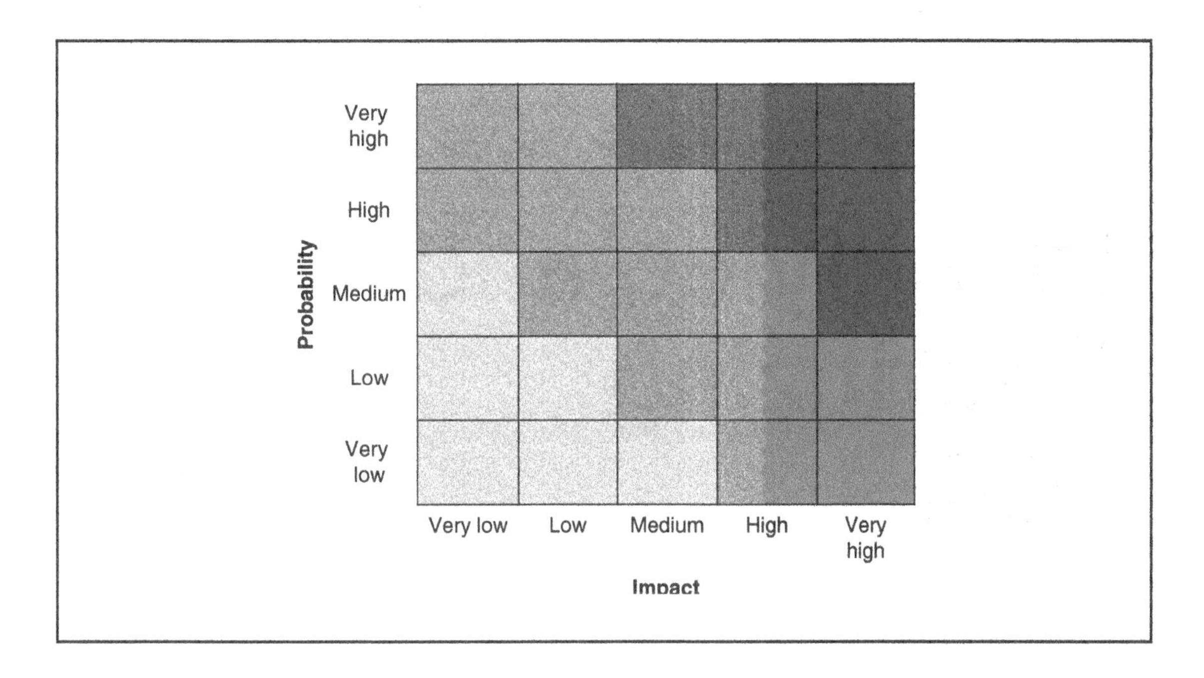

SUMMARY

In this lesson we defined key terms and concepts associated with project risk management. We differentiated between a risk, issue, and impediment, as well as describing known and unknown risks. We looked at the role risk tolerance plays in establishing risk thresholds and creating probability and impact matrixes.

The elements in the risk management plan were described, and we showed a sample risk management plan for the Dionysus Winery.

Key Terms

impediment
issue
known risk
opportunity
risk
risk management plan
risk threshold
risk tolerance
threat
unknown risk

15

Identifying and Prioritizing Risk

Identifying risks is everyone's job. Even though the risk management plan defines roles and responsibilities for risk management, everyone should be on the lookout for events that could negatively affect the project as well as opportunities to improve project performance. Once risks are identified, they need to be analyzed and prioritized.

In this chapter you will learn several methods for identifying risks. In addition to the standard practice of analyzing risk by evaluating probability and impact, we will look at several additional variables you can use to analyze risk. You can use these variables to calculate a risk score. Then you will be able to prioritize risk based on the risk score.

Some risks may require additional analysis, so we will go over two ways to quantify risks.

IDENTIFYING RISKS

Risk identification and the subsequent analysis and response take place throughout the project. The project charter usually identifies high-level risks. These are just broad topics, not risk events. For example, the Dionysus Winery project identified these four areas as having a lot of uncertainty:

- Availability of staff with needed skills;
- Timeliness of permits and certificates of occupancy;
- Availability of materials;
- Cost of materials.

As you can see, these topics are too broad to develop responses; however, they do provide some initial areas you can focus on when starting to identify risk events.

Identification Methods

It's challenging to sit down with a blank sheet of paper and start identifying risks. To make it easier, there are several methods you can use to develop a robust list of risks. One of the easiest ways to begin identifying risks is to talk with your team and other stakeholders. You can brainstorm with your team, interview subject matter experts, and talk with people who have worked on similar projects.

Brainstorming: When you are brainstorming to identify risks, it is helpful to provide the team with some prompts or risk categories. For example, the Dionysus Winery project uses the following four categories as prompts:

- Construction;
- Regulatory;
- Supply chain;
- Staffing.

There are many ways to conduct a brainstorming session. One that can be useful for risk identification is to follow the steps below:

1. In a meeting room use one wall for each prompt;
2. Give each member of the team a set of sticky notes;
3. Divide the team into four groups, and assign each group to a prompt;
4. Give team members five minutes to write risks associated with that prompt on their stickies and post them on the wall;
5. At the end of five minutes, rotate the teams to the right; do this until each team has visited each prompt;
6. Review all the stickies under each prompt, and arrange them into subcategories (like an affinity diagram);

7. Remove duplicates;
8. Give team members time to review all the work and ask if there are any additional risks they would like to add under the subcategories.

When the team feels they have identified all the risks, collect the stickies and use them to populate a risk register.

Interviews: Interviews can be one-to-one or in small groups. Interviews are great for asking open-ended questions, such as:

If you were leading this project, what would you be most concerned about?
What risks can you think of associated with staffing?

Document analysis: Document analysis includes a review of all the aspects of the project management plan and the project documents to analyze, compare, and contrast the documentation. If you see any inconsistencies, record that as a risk. For example, if your schedule shows an independent vendor performing product testing but your budget doesn't account for the associated costs, you have an inconsistency. This will cause you either to overrun your budget or perhaps to not have the quality review by a third party that you need to ensure an unbiased assessment of the product. If the plans and documents that you review are incomplete or of poor quality, that is also a risk.

Analogous: Analogous risk identification entails comparing the current project to a prior similar project. You should review at least the risk register, schedule, budget, and lessons learned documents from previous projects. These documents can help you avoid repeating mistakes.

Assumptions and constraints: Your assumption log should be reviewed to identify what would happen if an assumption was not valid. For example, if you assume a sufficient local labor force to fill all the positions and the current labor market is tight, you might identify that as a risk. Constraints that are too aggressive, such as a schedule with no room for slippage or a fixed budget with very little reserve, could be a risk as well.

Checklists: Checklists should be used with caution. If you use them, make sure they are broad-based checklists, rather than yes/no checklists. For example, a broad-based checklist could state, *"Review all requirements for feasibility"* or *"Assess risk registers from previous similar projects."* These are better options than *"Has the sponsor signed off on the budget?"* or *"Have you identified the critical path?"*

Risk breakdown structure (RBS): A risk breakdown structure is a good tool to organize your thought process for identifying risks. You can use the categories identified in the charter as the highest level. Then decompose those into lower levels. For the Dionysus Winery the

Risk breakdown structure: A hierarchy of potential sources of risks.

categories are construction, regulatory, supply chain, and staffing. For the supply chain category, you might decompose the risk categories like this:

Supply chain:

1. Equipment
   - 1.1. Cost
   - 1.2. Schedule
   - 1.3. Availability
2. Material
   - 1.4. Cost
   - 1.5. Schedule
   - 1.6. Availability
3. Supplies
   - 1.7. Cost
   - 1.8. Schedule
   - 1.9. Availability

Remember, these are potential sources of risk. They are not risks.

You can also categorize risks by objectives, such as performance, schedule, cost, or stakeholder satisfaction risks. These would then be further decomposed. You can use the RBS as a starting place for risk brainstorming as described earlier.

Documenting Risks

Risks are documented in a risk register or risk log. When documenting risks in the risk register, it is important to write a complete risk statement, rather than a vague phrase. For example, *"The budget is at risk"* does not provide accurate or meaningful information. There is very little information in that statement that you can use to develop an effective risk response. In contrast, this risk statement is better:

> **Risk register:** A log for documenting information on threats and opportunities.

"Because the cost of the outdoor security cameras is greater than anticipated, there is a risk we will overrun the budget for that work package."

A well-written risk statement starts with the cause or event and then defines the impact. In the risk statement above, the cause is the cost of the security cameras being greater than anticipated, and the impact is an overrun for the work package budget.

The following risk statement identifies an event rather than the cause of the risk:

"The city may not approve the plans for the restaurant,"

and the impact:

"causing a schedule delay."

It is useful to add some context about the risk in a section for comments. Table 15-1 shows a risk ID, risk statement, and comments for the Dionysus Winery project.

Note: The risk register and risk management plan cover different aspects of risk. The risk management plan describes how risk management will be structured and performed on the project. It is a management document. The risk register keeps track of identified risks, their ranking, responses, and all other information about individual risks. It is a tracking document.

TABLE 15-1 Dionysus Winery Risk Statements and Comments

ID	Risk Statement	Comments
1_T	Because the cost of the outdoor security cameras is greater than anticipated, there is a risk we will overrun the budget for that work package.	The cameras have gone up in price. We haven't been able to find any at a lower rate. The overall cost impact is $1000. This is a minimal impact to the overall budget and can be funded by contingency reserve.
2_T	The city may not approve the plans for the restaurant, causing a schedule delay.	If the city does not approve the plans as submitted, we will have to revise them. To stay on schedule, we will need to fast-track some activities and crash others. Thus, the schedule and budget will be affected.
3_T	Because of a labor shortage for housekeeping staff, we may not be able to staff all the positions, causing a degradation in performance for the hotel.	Availability of service staff is limited, and there is significant competition. Potential candidates are not top quality. Potential impacts include poor customer service, limited housekeeping, and operating at reduced capacity.
4_T	Due to global supply chain issues, the de-stemmer crusher may be delayed, causing a delay in the start of operations for the wine production facility.	The de-stemmer crusher is manufactured in Italy. There is currently about a four-month wait for equipment shipped from Europe.
5_O	Because a nearby winery is closing its doors, there is an opportunity to acquire some of their wine barrels at a discounted rate.	Newer barrels provide more tannins and thus a more robust taste. The used barrels are 50% less than the new barrels.

Legend: T = Threat, O = Opportunity

ANALYZING AND PRIORITIZING RISKS

One of the most common ways to analyze risks is to determine the probability and impact for each risk event. You will use the definitions of probability and impact that were documented in the risk management plan.

Filling Out the Probability and Impact Matrix

Once you have determined the probability and impact, you can map the results on a probability and impact matrix. This provides a visual representation of the estimated likelihood of each event occurring and the impact if the event does occur. When you multiply the probability times the impact you arrive at a risk score. The risk score is what you will use to rank the risks. Table 15-2 shows the risk register for the Dionysus Winery with the probability and impact information added. In a P x I matrix, lower numbers equate to lower probability or impact, higher numbers equate to a higher probability or impact.

Since we are evaluating multiple impacts, the risk score is the sum of the probability times the impact for each event. For example, for risk 2, the probability is 2, the impact

TABLE 15-2 Dionysus Winery Probability and Impact Analysis

ID	Risk Statement	Probability	Impact			Score
			Time	Cost	Performance	
1_T	Because the cost of the outdoor security cameras is greater than anticipated, we may overrun the budget for that work package.	5	0	2	0	10
2_T	The city may not approve the plans for the restaurant, causing a schedule delay.	2	4	3	0	14
3_T	Because of a labor shortage for housekeeping staff, we may not be able to staff all the positions, causing a degradation in performance for the hotel.	3	0	0	4	12
4_T	Due to global supply chain issues, the destemmer crusher may be delayed, causing a delay in the start of operations for the wine production facility.	5	5	0	4	45
5_O	Because a nearby winery is closing its doors, there is an opportunity to purchase some of their wine barrels at a discounted rate.	3	0	4	3	21

on time is 4 and the impact on cost is 3. You would calculate the risk score like this: (2 × 4)+(2 × 3) = 14.

The probability and impact matrix for performance is shown in Figure 15-1. If you are evaluating the impact of each objective separately, you would have a separate P x I matrix for cost and a separate one for schedule.

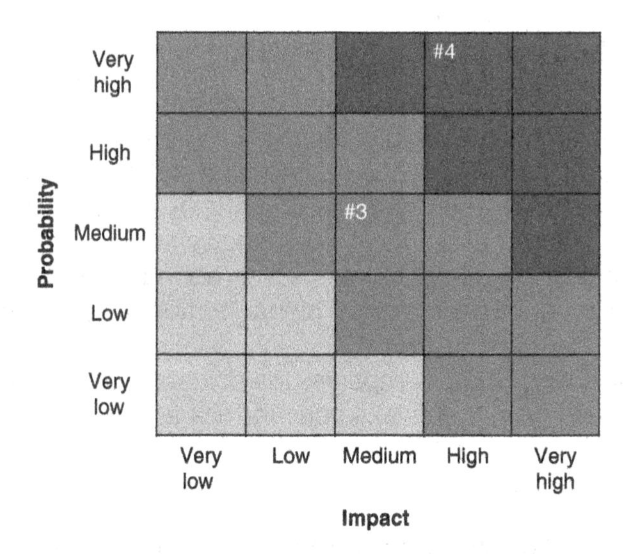

FIGURE 15-1 Probability and impact matrix for performance threats.

Opportunities are mapped on a similar but separate matrix. The difference is that a high impact score for an opportunity is favorable, while a high impact score for a threat is not.

Assessing Additional Risk Parameters

While probability and impact are the most common ways of assessing risk, they are not the only ones. For example, you may want to assess the urgency of the risk. In other words, determine how soon the response needs to be implemented in order to be effective. The following parameters are useful to consider when assessing risks:

Urgency: the immediacy with which the response needs to be implemented in order to be effective;
Proximity: how soon the risk is likely to occur;
Manageability: how easy it is to manage the risk event;
Controllability: how easy it is to control the outcome of the event;
Detectability: how easy it is to determine that the risk event has occurred;
Connectivity: how related the risk event is to other risk events;
Impact on organizational strategy: the degree to which the risk event affects the organization's strategy.

TABLE 15-3 Chart with Probability, Impact, and Urgency

ID	Risk Statement	Probability	Impact*	Urgency
1 T	Because the cost of the outdoor security cameras is greater than anticipated, we may overrun the budget for that work package.	5	2	1
2 T	The city may not approve the plans for the restaurant, causing a schedule delay.	2	7	3
3 T	Because of a labor shortage for housekeeping staff, we may not be able to staff all the positions, causing a degradation in performance for the hotel.	3	4	3
4 T	Due to global supply chain issues, the de-stemmer crusher may be delayed, causing a delay in the start of operations for the wine production facility.	5	9	5

* In this chart the impacts have been summed to get an overview.

Probability and impact charts only show two parameters. You can add a third parameter such as urgency, as shown in Table 15-3.

You can then create a bubble chart, like the one in Figure 15-2, to show the probability, impact, and urgency of a risk. The urgency is reflected by the size of the bubble.

Risk 4 is the top priority as it has a probability of 5, an impact of 9, and an urgency of 5.

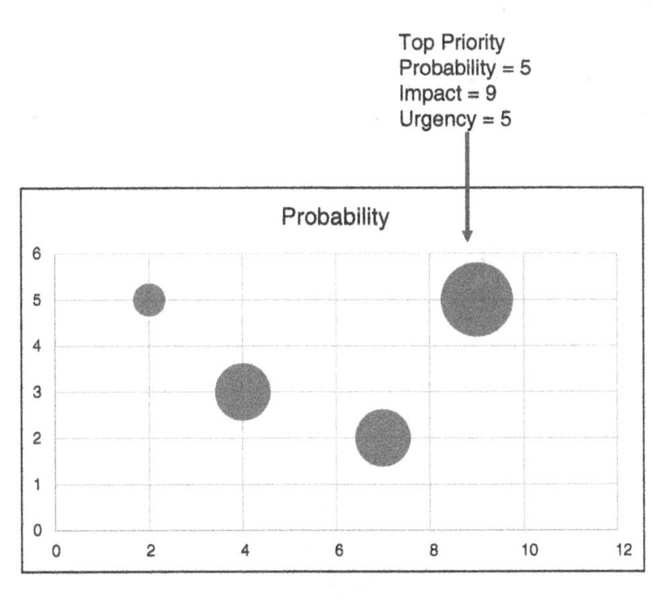

FIGURE 15-2 Bubble chart with probability, impact, and urgency.

SIMPLE QUANTITATIVE ANALYSIS METHODS

There are some risks that you will want to delve into a little deeper to understand the potential impact of a risk event. For those risks, you can employ some quantitative analysis techniques. Two of the simpler options for quantitative analysis are conducting an expected monetary value analysis and developing a decision tree.

Expected Monetary Value

Expected monetary value (EMV): A statistical technique that calculates the average outcome when the future includes multiple potential outcomes.

An expected monetary value (EMV) technique is useful when the impact of a risk is not well defined. It is a way of quantifying the uncertainty associated with multiple potential outcomes. As an example, we'll look at risk 2 for the Dionysus Winery: *The city may not approve the plans for the restaurant, causing a schedule delay.*

Tony Dakota is in the process of interviewing architects for the restaurant. One of the architects, Bailey Brothers, has a good price for the plans, $30,000. Based on their experience, the city accepts the plans as submitted 40% of the time. Sixty percent of the time the city sends the plans back for rework and resubmission. When this occurs, it adds 30 days to the process and $10,000 to the price.

Another architect, Avery Architecture, costs $33,000. However, the city approves their plans 70% of the time. For the 30% of the time they don't, it adds 30 days but only costs an additional $5000.

To develop the expected monetary value (EMV), start by putting this information in a chart, like the one shown in Table 15-4.

Next you will multiply the cost times the probability for each potential outcome, as shown in Table 15-5.

TABLE 15-4 Expected Monetary Value Setup

	Cost	Probability
Bailey Brothers		
Accepted	30,000	40%
Rejected	40,000	60%
Avery Architecture		
Accepted	33,000	70%
Rejected	38,000	30%

TABLE 15-5 Expected Monetary Value Path Calculation

	Cost	Probability	EMV
Bailey Brothers			
Accepted	30,000	40%	12,000
Rejected	40,000	60%	24,000
Avery Architecture			
Accepted	33,000	70%	23,100
Rejected	38,000	30%	11,400

To determine the EMV for each architect, sum the EMV values for the accepted and rejected costs. Thus, the expected monetary value for Bailey Brothers is $12,000 + $24,000 = $36,000. The expected monetary value for Avery Architecture is $23,100 + $11,400 = $34,500. Thus, Avery Architecture is the better choice because the expected cost is lower.

Given that you don't know if the city will or won't accept the plans, you can use this method to get an average outcome given the uncertainty of the situation. Here is a summary of the expected monetary value method.

1. Identify the outcomes that could occur;
2. Determine the probability of each outcome. Make sure the probability for each option adds up to 100%;
3. Determine the monetary value associated with each outcome;
4. Multiply the probability times the cost of each outcome;
5. Sum the costs for each option to get the EMV.

Expected monetary value can be used in other situations, not just risk. You can use this method for making decisions, looking at investment options, or other situations where you have multiple options and uncertainty around the outcomes.

Decision Trees

A decision tree uses the information from an expected monetary value calculation and turns it into a graphic. Figure 15-3 shows the same information as the EMV chart; however, it is laid out in a graphical format. A decision tree is a simple way to get a visual representation of the uncertainty and the various decision options for project risks.

Decision tree analysis: A diagramming technique for evaluating multiple options in the presence of uncertainty.

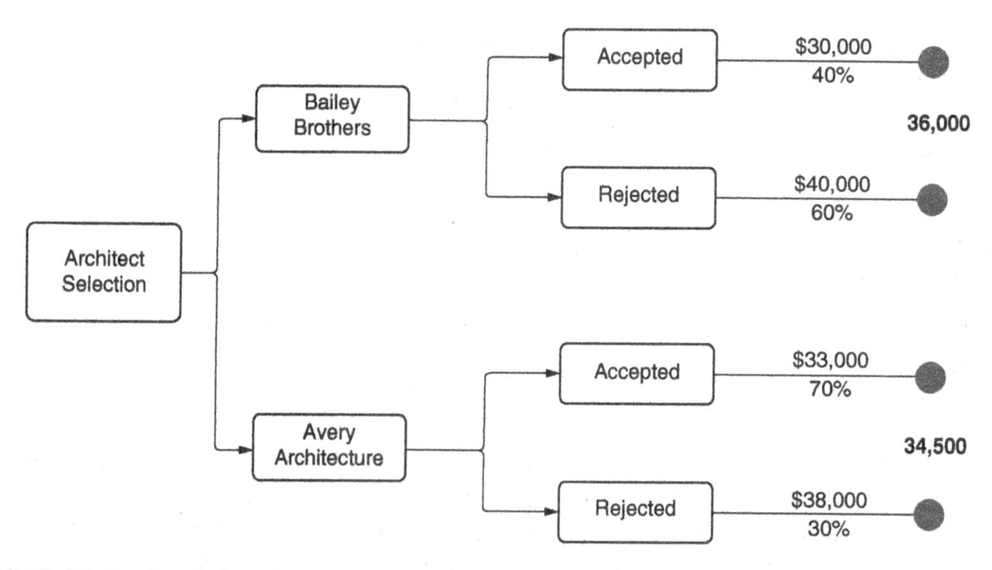

FIGURE 15-3 Decision tree.

You can show as many branches as you like on a decision tree; however, after about three main branches, it can get a bit complicated.

SUMMARY

In this lesson we introduced seven methods for identifying risk. We discussed the necessity of writing well-crafted risk statements in the risk register. Once risks are identified, they can be analyzed. We looked at how to apply the probability and impact definitions that are documented in the risk management plan and then plot them on a probability and impact matrix. This lesson described seven additional parameters you can use to analyze risk, and we showed a bubble chart that graphed probability, impact, and urgency. We finished the chapter by describing how you can use expected monetary value and a decision tree to quantify uncertainty.

Key Terms

decision tree analysis
expected monetary value
risk breakdown structure
risk register

16

Reducing Risk

Risk identification and analysis is necessary, but it is meaningless without following through by implementing risk responses. There are several ways you can respond to risks, ranging from avoiding them altogether to accepting them. For risks that you accept, you may choose to have a contingency plan or use contingency reserve.

In this chapter you will learn about five categories of risk responses that can be used for threats on predictive and adaptive projects. You will see how you can adjust a backlog to accommodate risk responses. We will also demonstrate different techniques for estimating the necessary contingency reserve to conduct risk activities and fund risks that you decide to accept.

RISK RESPONSES

There are some risks where you know right away how you should respond. Others require a bit more thought to determine the best response. In this chapter we will explore five categories of risk responses:

- Avoid;
- Mitigate;
- Transfer;
- Escalate; and
- Accept.

You can employ multiple responses to a risk event if you feel it is necessary. Threats and opportunities have similar responses; however, we will focus on responses for threats, with only a brief explanation of the complementary opportunity responses.

Risk Avoidance

Risk avoidance: Taking action to eliminate the threat.

When avoiding a risk, you're taking actions that eliminate the threat. Here are some options for avoiding threats.

- **Eliminate uncertainty:** If you have uncertainty associated with a deliverable, you can do more research to eliminate the uncertainty.
- **Relax or eliminate a constraint:** If you have multiple schedule risks, you can extend the schedule to avoid the schedule constraint that is putting the delivery date at risk.
- **Change a deliverable:** For risks associated with technology, you might be able to find a proven or less risky technology.

Secondary risk: A risk that occurs as a result of a risk response.

For all risk responses, but especially when avoiding risks, we sometimes introduce a new risk based on the response. This is called a secondary risk. Secondary risks must be entered into the risk register, analyzed and responded to like any other risk. The complement to avoiding a threat is exploiting an opportunity. Exploiting means taking actions to ensure you capture the opportunity.

Risk Mitigation

Risk mitigation: Reducing the probability and/or impact of a threat.

Mitigating a risk means finding ways to reduce the probability of an event occurring and/or the impact if it does occur. Options for mitigating threats include:

- Developing prototypes or models of products;
- Running additional tests; and
- Having redundant systems.

The complement of mitigating a threat is enhancing an opportunity. Enhancing entails taking steps to increase either the size of the opportunity or the likelihood of obtaining it.

Risk Transference

Risk transfer: Shifting the management and response of a threat to a third party.

Transferring risk usually equates to spending money to shift the risk management to another party. The two most common means of transference are insurance and contracts.

- **Insurance.** With an insurance policy you pay an insurer to absorb the financial risk of an uncertain event. However, be aware that if the event occurs, it is still your problem, the insurance company merely pays the bill.
- **Contracts:** You can transfer the work and the management of the risk to a vendor who might be better qualified to manage the risk. However, you are still accountable for the result, so if the vendor isn't effective, you could end up having to deal with the consequences of negative impacts.

Risk transference is another example where secondary risks associated with vendors can occur.

The complement of transferring a threat is sharing an opportunity. Sometimes by partnering with another party you can increase the likelihood of capturing an opportunity.

Risk Escalation

Escalation is used when the risk event is outside the scope of the project, or when you don't have the authority to implement the response. Examples of when escalation is an appropriate response include:

> **Risk escalation:** Bringing the risk to someone with more authority to address it.

- When the project is part of a program, and the threat could affect several projects in the program;
- If the response could affect departments or divisions that are not involved with the project; and
- If the cost of the threat response exceeds the project manager's budget authority.

Threats are usually escalated to the sponsor, PMO, program manager, or product owner. Risk escalation is the same regardless of whether you are responding to a threat or an opportunity.

Risk Acceptance

Risk acceptance can be passive or active. With passive acceptance if the event occurs, you deal with it in the moment. A common acceptance strategy is to have reserve time in the schedule or reserve funds in the budget to absorb overruns for unknown events.

> **Risk acceptance:** Acknowledging the existence of a risk but not taking action unless the risk occurs.

Risk trigger: An event or condition that indicates a risk has or is about to occur.

Fallback plan: A risk response used when other responses have failed.

You can also take a more active approach to acceptance, such as developing a contingency plan to implement in the event the risk materializes. When you have a contingency plan, it is useful to identify a risk trigger that lets you know that the risk is imminent.

In some cases, you may want to develop a fallback plan in case other risk responses are not effective. For example, if you are upgrading a system and it is causing a lot of problems, you can fall back on the old system.

Passive and active acceptance operate the same for both threats and opportunities.

IMPLEMENTING RESPONSES

Once you have identified risk responses, you should reanalyze your risks. You would expect to see the risk score drop based on your responses. Table 16-1 shows an example of the updated risk register for the Dionysus Winery once the risk responses have been determined. You can see a new row has been inserted under the probability and impact and score row. The new row is shaded and shows the change in rating after the response has been developed.

Note that the scores for risks 2, 3, and 4 went down. The score for risk 1 stayed the same because the team accepted the risk. Because risk 5 is an opportunity, the score went up due to the response.

When you have agreement on the risk responses, you will likely need to update various plans and project documents. For example, for the risks identified for the Dionysus Winery you would take the following steps.

Risk 1. No action required other than purchasing the cameras.

Risk 2. Assuming the risk analysis is done before baselining, the budget would reflect the higher cost for the vendor.

Risk 3. Work with the contracting department to bring a staffing agency on board. This may result in a secondary risk of increased costs, both for the staffing agency fees and the signing bonuses. You would likely escalate this to the project sponsor to confirm that performance is a higher priority than budget for this risk.

Risk 4. Update the schedule to order the de-stemmer crusher as soon as possible. You may incur storage fees if you don't have the space to store the equipment before the production facility is renovated.

Risk 5. Work with the relevant party at the nearby winery to purchase their barrels.

TABLE 16-1 Dionysus Winery Risk Register with Responses

			Impact			Score	Response
ID	**Risk Statement**	**Prob**	**Time**	**Cost**	**Performance**		
1_T	Because the cost of outdoor security cameras is greater than anticipated, we may overrun the budget for that work package.	5	0	2	0	10	**Accept:** Use contingency funds.
		5	0	2	0	10	
2_T	The city may not approve the plans for the restaurant, causing a schedule delay.	2	4	3	0	14	**Mitigate:** Pay more for a vendor with a better track record of getting city approval.
		1	4	3	0	7	
3_T	Because of a labor shortage for housekeeping staff, we may not be able to staff all the positions, causing a degradation in performance for the hotel.	3	0	0	4	12	**Transfer:** Work with a staffing agency. **Mitigate:** Offer a $500 signing bonus.
		1	0	0	2	2	
4_T	Due to global supply chain issues, the de-stemmer crusher may be delayed, causing a delay in the start of operations for the wine production facility.	5	5	0	4	45	**Avoid:** Order the equipment at the start of the project. This allows 11 months for equipment that may incur a 4-month delay.
		0	0	0	0	0	
5_O	Because a nearby winery is closing its doors, there is an opportunity to acquire some of their wine barrels at a discounted rate.	3	0	4	3	21	**Exploit:** Purchase the barrels.
		5	0	4	3	35	

RISK-ADJUSTED BACKLOG

Projects that use Agile methods may keep a risk register or impediment log. One of the principles associated with Agile is early and continuous delivery. Thus, anything that gets in the way of delivery is a risk or impediment. Project work is kept in a backlog that can be reprioritized at the start of each iteration. Therefore,

Risk-adjusted backlog: A backlog that incorporates actions to reduce impediments or threats to the project.

the product owner may choose to incorporate risk reduction work items into the backlog and prioritize them along with the development work. This results in a risk-adjusted backlog.

Here are three sample risks from the Dionysus Winery management system:

1. Greek God Wineries is updating their enterprise architecture. There are two options under consideration; one will not entail a change to our interface mapping, the other will.
2. Greek God Wineries is installing a new knowledge management system at the same time we will be creating the user instructions. The system will not allow new content until after the new knowledge management system is installed and tested. This will cause a delay in user instructions.
3. The cloud carrier is upgrading their system. Once the upgraded system is live, we will have to test our data and functionality to ensure it works correctly in the new environment.

These risks will need to be addressed and the responses incorporated into the prioritized backlog.

Figure 16-1 assumes the first iteration is complete. At the iteration planning meeting the product owner reprioritized the backlog. The following changes were made:

1. A new task was added to develop an interface map that would meet the needs of the other potential enterprise architecture. The team estimated that it would take less effort than the original interface mapping since they could reuse some of the work.
2. Writing the user instructions was moved up to accommodate the new knowledge management system. By getting the work done early, the user instructions will be brought over when the new knowledge system cuts over.
3. A new task was added to test the upgraded cloud environment. It is scheduled for the next iteration.

The new or reordered tasks are shown with a thick border.

For hybrid projects you would track all risks and responses on the risk register to ensure you have an overarching view of risk for the project. As you can see, these risk responses added another iteration to the schedule. The project manager and product owner would need to confer to determine if additional team members could be brought in to clean the data and thus shorten the duration or if the few extra days can be absorbed in the overall schedule.

Tasks	Points	Iteration
Review enterprise architecture	3	1
Review enterprise architecture policies	3	1
Map interfaces	8	1
Develop process diagram	21	1
Develop backup interface map	5	2
Identify data sets	3	2
Build data model	13	2
Write user instructions	13	2
Test upgraded cloud environment	5	3
Build prototype	21	3
Import data	8	3
Clean data	13	4

FIGURE 16-1 Risk-adjusted backlog.

RESERVE

Reserve is an umbrella term that refers to additional funds or additional time that the project team uses to deal with risk. There are two types of reserve, contingency reserve and management reserve.

Reserve: Additional funds or time added to a baseline to account for cost or schedule risk.

Contingency Reserve

Contingency reserve is usually under the control of the project manager. It is used to accommodate accepted risks and variances that are within the variance thresholds. Because contingency reserve is considered part of the cost

Contingency reserve: Funds or time incorporated into a baseline to accommodate known risks and risk responses.

and schedule baselines, it is expected to be used. This gives the project manager the freedom to make decisions for the good of the project, such as taking a few extra days to ensure the requirements are complete or spending some funds for a subject matter expert to provide feedback on a prototype. These decisions can ultimately reduce risk on the project and lead to better outcomes.

One of the easiest ways to determine the appropriate amount of reserve is to take a percentage of the schedule duration or budget. For low-risk projects, 10% is standard. For high-risk projects, 50% is more realistic. But these are very broad parameters. To get a more accurate number, you can develop risk-adjusted estimates.

Risk-Adjusted Estimates

Risk-adjusted estimates consider the range of outcomes associated with a work package by using the multipoint estimating technique described in Chapter 10. This generally entails interviewing the people who will be doing the work and asking them what their optimistic estimate is for the work, their pessimistic estimate, and their most likely estimate. When doing this, it helps to understand the rationale for each estimate. In other words, what would have to happen for the optimistic estimate, what would go wrong for the pessimistic estimate, and what assumptions are present for the most likely estimate.

Table 16-2 shows five work packages with optimistic, most likely, and pessimistic cost estimates.

For risk-adjusted estimates, a beta distribution equation is used to determine an expected value. The beta distribution equation is $\frac{O+4M+P}{6}$. This equation weights the most likely estimate more than the optimistic and pessimistic estimates. The expected value is an average outcome. In other words, 50% of the time the results will be less than the expected value, and 50% of the time they will be greater than the expected value.

When you apply the weighted average equation and sum the totals, you can see the outcome in Table 16-3.

TABLE 16-2 Calculating Risk-Adjusted Estimates—Step 1

Work package	Optimistic	Most likely	Pessimistic
A	$ 2,000	$ 3,050	$ 5,000
B	$ 1,500	$ 1,600	$ 1,850
C	$ 3,000	$ 5,100	$ 6,000
D	$ 4,500	$ 5,200	$ 6,200
E	$ 2,500	$ 3,000	$ 4,100
Total	**$13,500**	**$17,950**	**$23,150**

TABLE 16-3 Calculating Risk-Adjusted Estimates—Step 2

Work package	Optimistic	Most likely	Pessimistic	Expected value
A	$ 2,000	$ 3,050	$ 5,000	$ 3,200
B	$ 1,500	$ 1,600	$ 1,850	$ 1,625
C	$ 3,000	$ 5,100	$ 6,000	$ 4,900
D	$ 4,500	$ 5,200	$ 6,200	$ 5,250
E	$ 2,500	$ 3,000	$ 4,100	$ 3,100
Total	**$13,500**	**$17,950**	**$23,150**	**$18,075**

If your expected value total is greater than the sum of your most likely estimates, there is less than 50% likelihood that you will achieve your cost or schedule goal. Which means the sum of the most likely estimates is not very likely!

Once you discover that the most likely estimates are not sufficient, what do you do? There are two ways to handle this; one uses a spreadsheet to provide a quick but not precisely accurate estimate, and the other entails modeling software, which provides excellent information for estimating and forecasting. Purchasing and using modeling software is usually considered overkill for projects that are less than $20,000,000, so we will focus on the simpler spreadsheet version.

Calculating Contingency Reserve

When you apply the beta distribution equation, you get a distribution of outcomes with the expected value as the 50% point. Figure 16-2 shows an outcome distribution with the sum of the most likely and the sum of the expected values indicated.

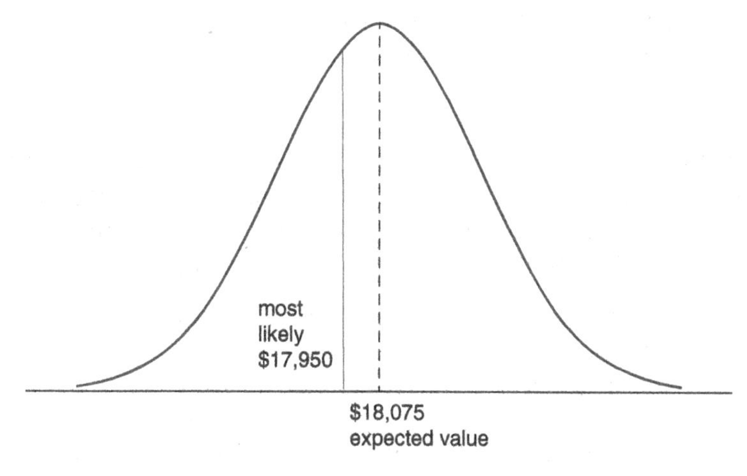

FIGURE 16-2 Calculating risk-adjusted estimates—step 3.

TABLE 16-4 Calculating Risk-Adjusted Estimates—Step 4

Work package	Optimistic	Most likely	Pessimistic	Expected value	Standard deviation	Variance
A	$ 2,000	$ 3,050	$ 5,000	$ 3,200	$ 500	$250,000
B	$ 1,500	$ 1,600	$ 1,850	$ 1,625	$ 58	$ 3,403
C	$ 3,000	$ 5,100	$ 6,000	$ 4,900	$ 500	$250,000
D	$ 4,500	$ 5,200	$ 6,200	$ 5,250	$ 283	$ 80,278
E	$ 2,500	$ 3,000	$ 4,100	$ 3,100	$ 267	$ 71,111
Total	**$13,500**	**$17,950**	**$23,150**	**$18,075**		**$654,792**

To get a rough estimate that will get you approximately in the 84% confidence range, subtract the optimistic estimates from the pessimistic estimates and divide by six for each work package. This gives you a standard deviation for each work package.

Next, square the standard deviation for each work package to get the variance. Table 16-4 shows the worksheet with the standard deviation and variance calculated.

Sum all the variances to get a total variance for the project; for our example the total variance is $654,792. Once you have the project variance, take the square root of that number. That is the standard deviation for the project. For our example, the standard deviation of the project is $809. If you add the standard deviation for the project to the sum of the expected values, you will get a number that you should be able to meet or beat 84% of the time. For our example the expected value plus one standard deviation is $18,884.

Tip: A best practice is to manage to the sum of the expected value and keep the standard deviation amount as contingency reserve.

Note: You may be wondering where all the equations come from, and why you have to divide everything by 6. These equations are based on probability theory and statistics. They represent outcomes if you are working with a normal distribution (bell curve). With a normal distribution, the following outcomes apply:

1. 50% of the results are greater than the mean (average), and 50% of the results are less than the mean.
2. The mean, median (midpoint), and mode (most frequent value) are the same value.
3. The standard deviation is specific to the measures for each work package, so you can't sum them.

4. 68% of the results fall within (±) one standard deviation of the mean.
5. 84% of the results are less than the mean + the standard deviation.

Management Reserve

Management reserve is different from contingency reserve. Management reserve covers unknown risks or unplanned, in-scope work.

Management reserve: Funds or time that is not part of a baseline. Used to address unplanned, in-scope work.

When projects are part of a larger program, we usually do not have access to management reserve. Management reserve is frequently held at the PMO, program, or portfolio level. Most smaller projects don't have a management reserve.

Examples of how management reserve might be used include:

- If you miss a requirement, management reserve would be used to pay for the unplanned in-scope work;
- If a component is failing and you need to rework it and conduct additional testing, management reserve would pay for the work and testing;
- If an unforeseen risk event occurs, management reserve covers the risk response or the impacts.

Tip: Risk management occurs throughout the project. Best practice is to do a thorough identification and analysis prior to baselining. This allows you to build in the risk responses into the project from the beginning. However, risks can surface at any time, even after you have baselined. Make sure you are constantly vigilant for new and evolving risks.

SUMMARY

In this chapter we looked at responding to risks. We covered five risk responses: avoid, mitigate, transfer, escalate, and accept. Some risk responses result in secondary risks, which need to be analyzed and responded to. For risks that are accepted, it is a good idea to establish risk triggers and fallback plans. Agile projects may integrate risk reduction activities into their backlog to create a risk-adjusted backlog.

Contingency reserve protects your cost and schedule baselines and allows you to make good decisions to reduce risk. We showed how to estimate contingency reserve by comparing most likely estimates to expected values and then calculating the needed contingency.

Key Terms

- contingency reserve
- fallback plan
- management reserve
- reserve
- risk acceptance
- risk avoidance
- risk escalation
- risk mitigation
- risk transference
- risk trigger
- risk-adjusted backlog
- secondary risk

17

Leading the Team

Throughout the project, one of our primary roles is to support the team in getting work done. This means making sure they have a healthy environment to work in and that they have a healthy mindset to accomplish the work. We also support the team by providing a structure to make decisions effectively and by making decisions fairly and in an unbiased way.

In this chapter we will discuss how to establish and maintain a psychologically safe environment where people feel comfortable expressing themselves. We will discuss how to foster qualities of adaptability and resilience in your team members. Because being a leader requires fair and unbiased thinking, we will discuss critical thinking and three of the main types of bias.

A lot of our day-to-day job entails solving problems and making decisions. Occasionally we are called on to resolve conflicts. We will describe techniques for each of these and provide some options you can use. With a shift toward remote work, it is likely you will have one or more remote team members. Therefore, we will discuss ways you can effectively work with team members in different locations.

ESTABLISHING A HEALTHY ENVIRONMENT

Throughout the project you will need to pay attention to the working environment, not just the physical environment, but also the team's well-being. This includes making sure that the environment is psychologically safe, that there is culture of adaptability, and that the team demonstrates resilience.

Psychological Safety

A psychologically safe environment is one in which people are comfortable expressing themselves without fear of retribution or embarrassment. It is a place where people feel free to be their authentic self. A psychologically safe environment is an inclusive environment where team members are valued regardless of religion, race, or sexual orientation. Diversity is valued and leveraged for the good of the whole.

Benefits of a Psychologically Safe Environment

Obviously it is beneficial to have an environment where we feel safe, and our team members do as well. Here are some of the many benefits of a psychologically safe work environment.

- **Improved feelings of trust:** When there is no fear or anxiety, there is more trust.
- **Higher morale:** People feel better about coming to work, and they feel good about the work they do.
- **Increased innovation:** Because people aren't afraid to fail, they can invent and innovate. This can also lead to new and more efficient ways of working.
- **Improved cooperation and collaboration:** When people feel safe, they are more likely to cooperate with each other and share ideas and knowledge collaboratively.
- **Reduced stress:** There is enough stress in life, so work shouldn't add to it. A psychologically safe environment reduces stress and anxiety.
- **Learning from mistakes:** When mistakes are seen as opportunities to improve, the organization learns. When mistakes are seen as bad, people tend to hide them, and they are often repeated.
- **Lower conflict:** With the appreciation of diverse opinions, there is less conflict.
- **Increased productivity:** All the above benefits lead to increased efficiency and productivity.

Creating a Safe Environment

As a project leader there are several steps you can take to create a safe environment. To begin with, talk to your team. Find out what is important to them in creating a healthy workplace. It is important that the team generates their own values so they have a sense of ownership. They may choose to document them in their team charter, or they may post them on the wall in a shared workspace. You can support them in operating consistently with their values as needed.

Encourage open conversation with and among your team. Your team will watch you to see whether you are authentic and whether your words and actions are aligned. Stay open to feedback and encourage honest and respectful communication. In addition to being open to feedback, establish a mechanism for people to express appreciation for one another. When people show and feel appreciation and acknowledgment, it increases the positive energy for the whole team.

When things don't go as planned, frame the situation as a learning opportunity. Considering something a failure reduces psychological safety and shuts down innovation. Look for opportunities for problem-solving, growth, and resiliency.

Make it easy to provide input, suggestions, and feedback. For example, you can include time in every meeting for suggestions and feedback, or you can have a physical or digital suggestion box.

Maintaining a Safe Environment

One of the most important ways you can maintain a safe work environment is by having zero tolerance for bullying or toxic behavior. Any tolerance can erase all the work you have put into creating a healthy environment.

Another key is in how you respond to feedback. In establishing a safe environment you seek to make it easy for people to provide feedback; however, if you ignore or judge the feedback, it won't be long before people stop providing any. There will be some feedback you act on, and some you don't. Regardless, make sure you respond in a way that is respectful and appreciative. Some tips to for receiving feedback in a positive way include:

Active inquiry: Active inquiry means adopting a curious attitude. Ask questions about feedback, processes, the team environment, and so forth.

Providing additional information: Sometimes you have information your team members don't. There may be times when you can't implement suggestions because of situations they have no knowledge of. As appropriate, share that knowledge so they can get a better understanding of the situation, and so they understand that their suggestions aren't being ignored.

Situational humility: We all know that managers and leaders don't have all the answers. A leader is a role on a team, they don't know everything, and they aren't always right. Make sure people know that it takes everyone's voice to make things work.

Cultivating Adaptability

Projects are all about change. They are introducing something new, improving processes, or any other manner of change. In addition to the change projects create, there are changes to the project, priorities, requirements, stakeholders, and so forth. That means that you and the team will need to cultivate the ability to adapt.

Adaptability isn't just about being able to adjust to new or changing conditions, environments, trends, and other circumstances; it is about being able to adjust quickly, calmly, and effectively. Change is often accompanied by stress, uncertainty, anxiety, self-doubt, and other limiting feelings. To move past these feelings and cultivate adaptability you can take steps to prepare for a changing environment. In this context, we aren't talking about a change, we're talking about updating the way you think about change.

Here are three ways you can prepare yourself and your team to function effectively in a rapidly changing environment.

1. **Be observant:** Rather than waiting for the next shift in the project, the competition, the market, and so forth, spend time observing what is happening. Look for trends and indicators of what is likely to happen. This behavior can put you in front of the change rather than being taken by surprise. Being ahead of the change allows your team to maintain a sense of calm and prepare for what is coming.
2. **Develop a growth mindset:** Rather than seeing change as something that is bad, frightening, or irritating, help your team focus on what you can all learn. What new skills can you develop? How will the new situation help? Find ways to turn the change to your advantage. This mindset will help everyone maintain a positive attitude and shift your thinking from victim to victor.
3. **Learn to accept change:** Things are going to change whether we want them to or not and whether we are ready for them to or not. Therefore, the faster your team can accept it, plan for it, and even leverage and grow from it, the happier everyone will be.

With change and transformation, preparation is only half the game; the other half is how you respond. There are several ways you can foster adaptability in the face of disruption. To start with, be curious and open-minded. Ask questions and listen with an open mind. Try and understand what led to the current situation, what it means for your team and your project, and how to support the change.

Next, think about the situation from multiple perspectives. Talk with your team members and colleagues. Get their take on the situation. Apply that curiosity mentioned above. When you can see a situation from multiple perspectives, you are more effective in dealing with the challenges it can bring.

Approach the situation as a problem-solving opportunity. There are plenty of problem-solving frameworks you can apply to provide some structure to the process. Most of them have these common elements:

> Define the problem → identify the solution criteria → generate options → evaluate the options using the criteria → choose the best solution → evaluate the result.

To strengthen an adaptability mindset, when generating options, look for innovative solutions, foster creative thinking, and stretch the imagination. Don't settle for the easiest or even the safest response. Think bigger, thing differently. You may end up with the easiest or safest response, but don't lose the opportunity for innovation.

Fostering Resilience

With projects we don't have the luxury of evolutionary change. We must adapt quickly and recover quickly. That is where resilience comes in. One of the best ways you can serve your team is by fostering resilience.

Here are four ways you can cultivate resilience for yourself and your team.

Resilience: The ability to adjust or recover readily from adversity, crisis, setbacks, change, and other significant sources of stress.

1. **Keep things in perspective:** While a disruption or change may seem like a major concern, if you can step back and look at it from a wider lens, you will often find it is not as monumental as you first thought. Keeping things in perspective can include asking yourself and your team: In the overall scheme of things, is this going to be a big deal? Or does it just seem that way now?
2. **Maintain a positive outlook:** Thinking of all the things that could go wrong, or how awful the situation is, is counterproductive. No matter what the situation, endeavor to find a way to maintain and model a positive attitude. The ability to recover from adversity is directly influenced by one's attitude. Pay attention to both internal words and external words. The things we tell ourselves are just as important as what we say out loud. Keep both conversations positive.
3. **Accept change:** Accepting change is a part of building resilience as well as adaptability. We can't recover and move on if we are still holding onto the past or wishing things were different. People who are resilient acknowledge what is and keep moving forward.
4. **Learn:** The most resilient people are always learning. We can learn from positive as well as negative outcomes. We can learn from peers, mentors, and friends. Spend time reflecting to see what behaviors or actions you can carry forward and which you should adjust in the future.

WAYS OF THINKING

The way we think is a huge contributor to project success. There is no room for sloppy thinking when you are leading a project. We can't let feelings and hunches lead the way—we need to think critically, and we need to be on the lookout for bias. Another aspect of thinking that we should pay attention to is the need to shift between strategic thinking, system thinking, and tactical thinking. We'll look at critical thinking, bias, and system thinking next.

Critical Thinking

Critical thinking is considering problems and situations using logic and reasoning. While this sounds easy, we are hampered by our personal experiences along with thought processes that are based in self-interest. There are several aspects of critical thinking that we will work through.

Figure 17-1 shows an image to help visualize moving toward being a critical thinker. The first thing to remember is that each person has their own frame of reference. Our frame of reference is composed of our experiences, which create beliefs, assumptions, bias, and even distortions in the way we interpret events around us. Because we are individuals, these things are unique to each of us. Given that we don't have the same frame of reference, when we listen to another person's assumptions or beliefs, they can seem illogical or irrational.

Based on the way we have evolved over millennia, we are egocentric, in other words, we think of ourselves first. We have our self-interest in mind, and we consider our social

Frame of Reference

Each of us has our own unique set of these

Beliefs
Assumptions
Experiences
Bias
Distortions

Self-interest
Egocentrism
Sociocentrism

Human nature leads us to think like this

How we think

FIGURE 17-1 Thinking as usual.

circles or cultures before considering others. This may only take a fraction of a second, but it is hard-wired into the way we think.

To think critically, we need to deliberately shift how we think. Rather than think in our own best interest, we should strive to be open-minded, fair minded, and consider alternatives. We should ask ourselves if we are being reasonable and rational in our thought process. This is shown in Figure 17-2.

Here is a four-step process to think more critically.

1. **Gather information:** Gathering relevant information may include conducting interviews, performing research, searching the Internet, holding focus groups, and so forth.
2. **Analyze the data:** Analyzing data begins with examining evidence uncovered in your research and evaluating arguments for and against various options. You can also look for logical relationships between variables. This can include evaluating if there is a correlation between variables, a cause-and-effect relationship, or no relationship.

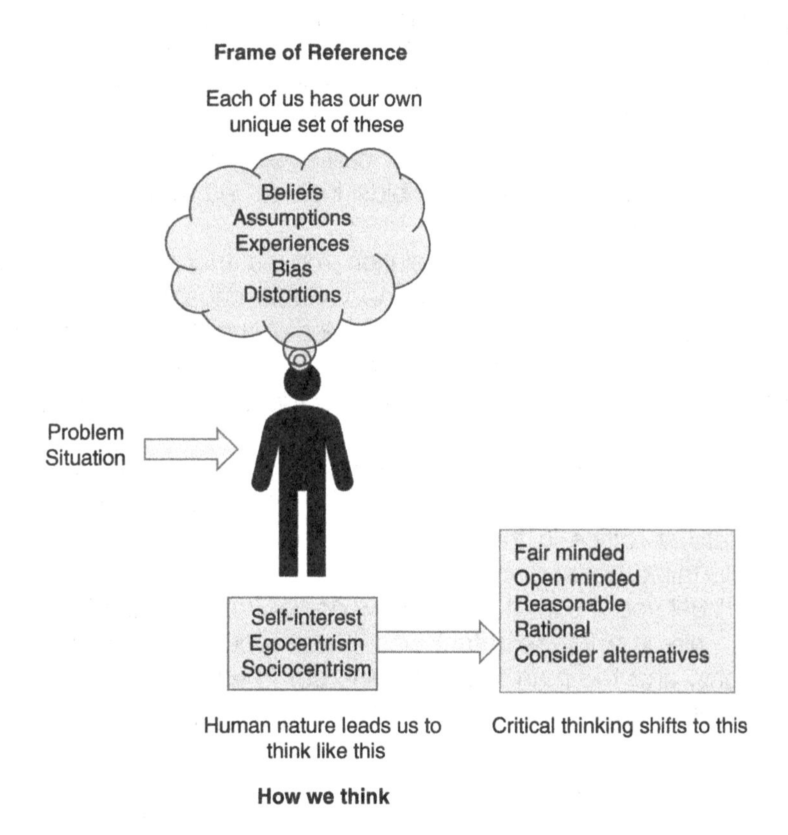

FIGURE 17-2 Being a critical thinker.

3. **Draw conclusions:** After a thorough analysis you will draw conclusions.
4. **Test conclusions against criteria or standards:** Before you finalize your position, evaluate it against unbiased criteria or standards. This can include professional standards, such as the Project Management Institute's Standard for Project Management, or de facto standards, such as best practices.

Working with Bias

As mentioned above, we each have our own set of biases. The tricky thing about biases is that they are rarely transparent to the people who hold them. Unfortunately, biases can often result in unfair treatment. For example, having a bias that someone with a college deegree is more qualified for a job than someone without one could lead you to ignore qualified candidates who have significant experience, but no degree. There are three types of biases that inhibit our ability to think critically: anchoring bias, correlation implies causation, and confirmation bias.

Anchoring: You've heard how important a first impression is. Anchoring is relying too much on our first impression or the first information we see. This can influence how we evaluate later information. In getting a bid for pouring a foundation, the first bid tends to be the anchor from which people evaluate other bids. If there are three bids, the first is $8,000, the second is $10,000 and the third is $9,500, a person might say that the second and third bids are too pricey. An anchor can even get in the way of seeing clearly, despite evidence, such as if subsequent bids include warranties, grading, or striping work, when the initial bid does not.

Correlation implies causation: Correlation means there is a relationship or pattern between variables. Causation means one variable causes the other. For example, data shows that in July and August, ice cream sales go up. Data also shows that in July and August more people get sunburned. There is a correlation between these two variables. But one does not cause the other. In fact, the cause for both is an increase in temperature.

Confirmation bias: Confirmation bias is interpreting information in such a way as to confirm your beliefs. It includes ignoring information that is inconsistent with your beliefs. Imagine walking into an office in the 1980s, and there is a man and a woman talking. Most people would think the woman was the secretary and the man was the boss. Even if the man said, "I'll get you a cup of coffee, how do you take it?" Confirmation bias would lead the person to assume the secretary wasn't doing her job. In fact, the woman may have been the boss, and the man the secretary, but it would be really hard to dislodge the confirmation bias.

System Thinking

> **System:** A set of interacting, interrelated, or interdependent elements, forming a whole.

If you think about it, you could see a project as a system. A project is a set of deliverables that interact, interrelate, and are interdependent. Taken together the deliverables form a whole. An organization is a system as well. Therefore, you could consider that projects are systems within systems. Following this line of reasoning, understanding systems thinking can help you approach your project with a holistic perspective.

One of the aspects that makes systems complex is adding stakeholders into the mix. Stakeholders have different needs, opinions, and biases. They affect operations, new product development, marketing, and so forth.

Let's look at an example of how a risk at the Dionysus Winery can be viewed using system thinking. In the previous chapter we identified the following risk: *Because of a labor shortage for housekeeping staff, we may not be able to staff all the positions, causing a degradation in performance for the hotel*. If we look at this risk from a system perspective, we can understand the context of the risk more and the implications of the proposed response.

Housekeeping is a part of maintaining the hotel. Fully staffed housekeeping allows for full occupancy, better guest experience, and better reviews on social media. Higher occupancy is one of the factors that increases restaurant receipts, winery tours, and wine club membership. Thus, we can see the interrelatedness and interdependency of housekeeping with other functions of the hotel and bottom line profitability.

The risk response was to transfer the work to a staffing agency and offer a $500 signing bonus. Looking at the staffing from a system perspective, offering a signing bonus for housekeeping, but not grounds maintenance, wait staff, and front desk staff could have unintended consequences of those roles finding out about the bonus and wanting a signing bonus as well. If they don't get one, they may look elsewhere, leading to a different set of staffing problems. Viewing the project from a system perspective helps to identify relationships between different parts of the project and the potential consequences of decisions. This helps us make better decisions for the overall project outcomes.

SUPPORTING THE TEAM

One of the things that creates delays on projects are problems that arise, especially when analyzing and generating potential solutions entails meetings, which take people away from work. The same can be said of decisions that need to be made. Many times no one wants to be accountable for making a decision, especially if there are potential negative repercussions. Conflicts can also take time away from productive work.

As project managers we can support the team's progress by guiding people through a problem-solving and decision-making process. We can also employ various methods of working through conflict to keep the team focused on the outcomes and the project moving forward.

Solving Problems

When faced with a problem, particularly big problems, it can be difficult to know where to start. The six-step process shown in Figure 17-3 provides a framework that will allow you to focus on the problem rather than figuring out a process for solving it.

FIGURE 17-3 Problem-solving process.

Define the problem: Many times, solving a problem is difficult because we aren't clear what the problem is that we are trying to solve. The first step is to make sure everyone involved is aligned with the problem definition. You should be able to clearly articulate the problem in a few concise sentences.

Identify solution criteria: Define the important factors in reaching a decision. Is time the driver? Is technical performance the most important element? You might want to set up a weighted scoring mechanism similar to the decision matrix described in Chapter 6.

Generate options: Make sure to consider various options as well as the implications and risks associated with each.

Evaluate the options: Define the pros and cons of each option and apply the weighting criteria (if used) to each alternative.

Choose a solution: Once you have evaluated the options, consider if there are any significant risks to the potential solution. Taking the risks into consideration, you can then choose the best solution.

Evaluate the result: You can evaluate the effectiveness of your problem-solving process and whether the resolution was effective. It may be useful to wait a few weeks in order to evaluate the results.

Making Decisions

You can apply the same process for solving problems to make decisions, but instead of defining the problem, you are defining the decision you need to make. There are different decision-making styles you can use. The style you use depends on a number of factors. Four decision making styles are:

- **Command:** This style is used when time is of the essence. It can also be used when the stakes are high. Be aware that it is not a good style if maintaining good relationships is important.

- **Consultation:** Consultation is used when you need input and information to help make an informed decision. One person is still making the decision, but he or she seeks input before making the decision.
- **Consensus:** Consensus is the best style when you need buy-in from the people involved. In some cases, the people involved in the decision will agree to a majority or plurality voting block to reach a decision.
- **Random:** Random is like a coin toss. It's used when any solution is fine.

The decision-making style you use will depend on the factors involved:

Time constraints: If time is of the essence, you might need to use a command or random method. Consultation and consensus take more time.
Trust: If you trust the people involved in making the decision, or if you need to build trust, you should use a consultation or consensus model.
Quality: Consensus decisions tend to lead to better decisions. Random decisions are least likely to lead to a good decision.
Acceptance: If you need to have acceptance, you are best served by using a consensus style. You might be able to use a consultative style and still gain acceptance.

Resolving Conflicts

Sometimes conflicts can be addressed using the same framework as making decisions. This might be appropriate if the conflict is the result of a decision that needs to be made. However, there are numerous opportunities for conflict on projects, such as:

- Schedule;
- Priorities between the objectives;
- Scarce resources;
- Technical approach; and
- Cost.

Occasionally you might see some conflicts over administrative issues or personalities, but these are less common. Depending on the nature of the conflict and how important it is to keep a favorable working relationship, you can choose between six conflict resolution strategies:

- Confronting/Problem-Solving;
- Collaborating;
- Compromising;
- Smoothing/Accommodating;
- Forcing/Competing; and
- Withdrawal/Avoiding.

Confronting/Problem-Solving

Confronting doesn't mean being confrontational—it means confronting the conflict as a problem to be solved. This method is most useful when the relationship is important and you have confidence in the other party's ability to problem-solve with you. The objective is to reach a win-win situation.

Collaborating

Collaborating involves meeting to resolve the issues that are causing the conflict. This is most useful when you trust the other party (or parties) and when you have time to dig into the issues. This method involves incorporating multiple views and learning what is important to each party. This can take some time, but relationships are maintained and collaborating usually ends up producing the best results.

Compromising

Compromising is used when you are looking for some degree of satisfaction for both parties. Both parties need to be willing to give and take. This method is often used when there is an equal relationship between parties and when each party needs a win. This method also keeps the conflict from escalating.

Smoothing/Accommodating

Smoothing, also called accommodating, involves emphasizing areas of agreement and de-emphasizing areas of conflict. Use this method if maintaining a harmonious relationship is as important or more important than resolving the conflict. This is also a good method to use when you are likely to lose the conflict anyway, as it helps maintain good relations with the other party.

Forcing/Competing

Forcing or competing is a win-lose proposition. You are imposing the resolution. This is not a terribly productive method, but it may be appropriate when the stakes are high and you have a limited time. You should only use this when you know you are right. Be aware that this method is not a good one if you want to maintain a productive relationship.

Withdrawal/Avoiding

This method entails retreating from the situation. You may be retreating because the issue will go away on its own or if you know you can't win. It can also be used as a cooling off period if tempers are flaring.

Table 17-1 summarizes the conflict resolution techniques and when to use them.

TABLE 17-1 Conflict Resolution Techniques

Confronting	Collaborating	Compromising	Smoothing	Forcing	Withdrawal
When you have confidence in the other party's ability to problem-solve When a relationship is important When you need a win-win solution	When there is time and trust When you want to incorporate multiple views When there is time to reach consensus	When there is a willingness to give and take When both parties need a win When an equal relationship exists To avoid a fight	To maintain harmony When you will lose anyway To create goodwill	When you are right When the stakes are high If the relationship is not important When time is of the essence	When you can't win When you need a cooling off period If the problem will go away on its own

CONSIDERATIONS FOR VIRTUAL TEAMS

Whether you have one virtual member, several, or you are leading a team where everyone works in different places, the dynamics for virtual teams are different than co-located teams. To be the most effective leader possible, you will need to take some extra steps to keep people engaged. You can put in a bit more structure than you would for teams that work together every day. Meetings will also be different with a partial or fully virtual team, so you'll need to think about virtual meetings differently than you do in-person meetings.

Engagement

There are some challenges with virtual teams that aren't as common on in-person teams. You will likely find it more difficult to establish rapport with team members, build a cohesive team, communicate effectively, maintain commitment, and avoid burnout. We'll look at some techniques to address each of these challenges.

Establishing rapport: It can be challenging to build rapport with people you don't meet with face-to-face. You'll need to take some extra time to get to know each person. Find out what motivates them. What do they like to do in their spare time? What do you have in common? Let team members get to know you as well. For example, share some of your personal and professional likes and dislikes. This effort builds a sense of connection between you and your team members.

You can also have a weekly virtual video coffee break. Take a half hour where people can grab a cup of coffee, share, and hang out. You might want to introduce themes, such as crazy t-shirt day or who has the most interesting coffee cup.

Building a cohesive team: You can build a sense of team by meeting more often, such as having a short 15-minute stand-up meeting each day to check on progress and see whether there are any potential problems. You can also set up a virtual collaborative team space. In addition to project artifacts, have people post pictures there, start chat threads, even post comics or jokes. Even though it is virtual, try and make it an engaging environment.

Communicating effectively: Communication can be more challenging on a virtual team. It's a lot easier to misunderstand people or miss essential information when you aren't meeting in person. You'll need to spend more time making sure you're being very clear. Ideas include creating a bullet point list of the things you want to get across before a meeting; that way you can make sure you cover everything. When communicating in writing, send a first draft out to a colleague to see whether you're being clear and complete.

Lack of communication can be a challenge as well. If a team member goes silent, make the extra effort to reach out to them. With a virtual environment you have no way of knowing whether they're not communicating because they have a sick child or they're hopelessly behind and don't want to admit it.

Lack of commitment: It's often harder to keep people engaged in a virtual environment. Some of this is a matter of "out of sight, out of mind." When you have people working from home, you aren't just competing with other work matters for their attention. You are also competing with household chores, family members, and many other daily distractions. You can help by keeping the purpose and vision of the project visible; post the vision statement in your collaboration space. This reminds people about the difference their work will make.

Burnout: When people work from home they can't escape work—they feel tethered to their computer. People tend to work longer hours. While that might be good for productivity, it can lead to burnout, you can't control the other demands on your team members' time, but you can make sure not to send emails or have meetings outside normal work hours. It also doesn't hurt to check in and see whether people are feeling overwhelmed.

Structure

To accommodate a virtual environment, you will need to be proactive and deliberate in providing a structure that supports the team evolving into an effective one. One way to do this is by using multiple communication channels. For example, if the team has a big deliverable coming up, check on progress toward deadlines earlier than usual. You might leave a chat message on a collaboration site reminding people about the due date or send a quick email as a follow-up to mentioning the upcoming deadline on a team phone call. You will need to evaluate the additional communication (and the perception that you are pestering people) with the need to keep project work front and center.

You can also facilitate the team setting up team norms around working time. This is not to say that everyone must work the same hours, but it's nice if the team can work together online for a few hours each week. This can be an open working session, time used for problem-solving, or for exchanging ideas.

You might also set up "office hours," where you dial into a meeting room and anyone who wants to can drop in and talk. This doesn't take the place of team meetings or one-on-ones, but it's nice for the team to know you have specific times set aside for questions, ideas, or just chatting.

When your team is first forming, have a team retrospective to see what's working, what isn't, and what the team can try to improve. In the beginning you may want to have a video conference for the retrospective. As your team gets accustomed to working together you may have the meeting every two weeks, or even move to a virtual retro where the team can leave notes in a collaboration space about what's working and what they want to try differently.

Virtual Meetings

It can be challenging to keep people engaged with in-person meetings, but virtual meetings are even more challenging. Virtual meetings require you to be more engaged with planning the meeting and more creative in running the meeting. Here are a few tips you can use to make your virtual meetings more engaging.

- Have video meetings to allow everyone to see each other. This makes it easier to connect.
- Take a few minutes to chat about things that aren't just about work. This strengthens the sense of relatedness and rapport.
- Join the meeting 10 minutes early and invite people to come early and hang out.
- At the start of the meeting check in by asking everyone to use three words that describe how they're doing. Ideally, this can go in a chat box, or you can ask people to state their three words. This starts the meeting with a personal touch without using a lot of time.
- Wait longer for responses. People need time to process, and there can be a lag due to the communication technology you're using.
- Manage participation by not letting people overshare or go off on a tangent.
- Sequence interaction. If you ask for input and several people want to talk, state who has the floor first, second, third, and so forth. This eliminates talking over each other.
- Ask people if they're done before moving on to the next person. Since you don't have the visual cues that let you know if someone is done speaking, you may need to check with them.
- Make sure people minimize distractions by shutting their door, silencing their phone, not checking email during the call, and remind people to mute themselves if they're not talking.

SUMMARY

In this chapter we discussed how you can support your team in getting work done. We described a psychologically safe environment, how to build one, and how to maintain one. We also discussed how you can support team members by cultivating adaptability and fostering resilience.

Teams trust their leaders when we demonstrate rigorous, fair, and unbiased thinking—in other words, critical thinking. We described different types of biases and how you can guard against them. System thinking focuses on interactions, interrelationships, and/or interdependent elements that form a whole. Both critical thinking and system thinking should be applied when solving problems and making decisions. In addition to these topics, we discussed several techniques to resolve conflict.

We finished the chapter by discussing challenges and potential solutions for virtual teams. We focused on team engagement, putting in some light structure, and holding effective virtual meetings.

Key Terms

bias
resilience
system

18 Maintaining Momentum

No matter how well you have planned your project or how nimble you are in addressing new requirements, events and situations will happen that impede your progress. Being prepared to address changes will help you and your team maintain momentum.

In this chapter we will look at the change management process. You'll see how changes affect predictive work and adaptive work. We'll also discuss the implication of adding scope to predictive and adaptive work. To help you keep the project organized and moving forward, we'll introduce a few logs and charts you can use.

WORKING WITH CHANGE

As in life, the only certainty in projects is change. Because change is sure to happen, it is best to be prepared. Think through how you want people to submit change requests, how you will evaluate them, who can approve them, how you will track the changes, and so forth. You can record this information in a change management plan.

Change management: A process for identifying, tracking, evaluating, and implementing modifications for deliverables, plans, or project documents.

Change Management Plan

A change management plan has four main parts:

- Change management approach;
- Definitions of change;
- Change control board; and
- Change management process.

Change Management Approach

The change management approach describes the degree of change control and how it will be integrated with other aspects of the project. In a hybrid project it is a good idea to indicate which deliverables are subject to change control and which are not. For example, in the Dionysus Winery project, the winery management system would not be subject to the same degree of change control as the construction aspects of the project.

Configuration management: A system used to identify and track individual items in a product or release.

This section may also discuss how change management and configuration management will work together. Some projects don't require configuration management, and some do. Projects with thousands of parts, such as building an airplane, might have change management subsidiary to configuration management. Otherwise, configuration management is often subsidiary to change management. The needs of the project should inform the relationship between the two.

Definitions of Change

Suppose you are ordering wine racks for the Dionysus Winery. When you were gathering requirements the production manager said he wanted 10 wine racks that could hold 200 bottles each. A week before you place the order, he says that two of the wine racks should be secured so you can lock up the wine.

Is that a change? The number of racks hasn't changed. The number of bottles hasn't changed. If it is a change, should it be recorded as a change in scope? A change in requirements? A change in cost? While there is no absolute correct answer, it is a good idea to think through what is considered a change and what is not.

A change will usually require a change request, assessment of the impact on baselines, evaluation by a change control board, and possibly an update to baselines or other project documents.

Change Control Board

> **Change control board:** A group that reviews, analyzes, approves, or rejects changes to the product or the project.

The change control board (CCB) receives change requests, evaluates them, and makes a decision about whether the change is accepted, rejected, or deferred until more information is available. When documenting information in the change management plan you will want to consider what changes the project manager has the authority to approve, which changes need to go to the change control board for consideration, and which changes the customer or sponsor need to approve.

Change Management Process

The change control plan usually identifies at least these four steps to the change control process:

1. **Change request submittal:** This section identifies the forms that need to be filled out and who they are submitted to.
2. **Change request review:** Describes how the CCB will evaluate the request, including analysis of impact on scope, schedule, cost, customer satisfaction, risk, and so forth.
3. **Change tracking:** Change tracking describes how a change request is tracked from submission to implementation or rejection.
4. **Change outcome:** Describes how approved and rejected change requests will be handled, including how baselines will be updated, how often they will be updated, and how changes to baselines will be communicated.

Table 18-1 summarizes the information in a change control plan.

Change Requests

For predictive aspects of the project, a change request can come from anyone at any time after the plan is baselined. Usually there is a form that people fill out to request a change because verbal change requests are subject to misinterpretation. A change request usually contains the following information:

Change requestor: The name of the person requesting the change.

Description of the proposed change: A detailed description of the change. There needs to be enough detail so the change control board can adequately determine the impact to various aspects of the project.

Justification for the proposed change: Either the benefit of implementing the change or the implications of not implementing the change.

TABLE 18-1 Change Management Plan Contents

Document Element	Description
Change management approach	Describes the degree of change control and how change control will integrate with other aspects of project management
Definitions of change	
• Schedule change	• Defines a schedule change versus a schedule revision; indicates when a schedule variance needs to go through the change control process to be re-baselined
• Budget change	• Defines a budget change versus a budget update. Indicates when a budget variance needs to go through the change control process to be re-baselined
• Scope change	• Defines a scope change versus progressive elaboration; indicates when a scope modification needs to go through the change control process
• Project changes	• Defines when changes to the project, updates to project management plans, or updates to project documents need to go through the change control process
Change control board	Identifies members of the change control board, the extent of their authority, and how frequently they meet
Change management process	
• Change request submittal	• Describes the process to submit change requests and any templates, policies or procedures that need to be used
• Change request review	• Describes the process used to review change requests, including analysis of impact on project objectives
• Change tracking	• Describes the process for tracking change requests from submittal to final disposition
• Change outcome	• Describes possible outcomes such as accept, defer, or reject

Impact: Usually the impacts for a change are broken down by scope, quality, schedule, cost, and stakeholder satisfaction. You can include other impacts as well, such as an increased or decreased risk.

Comments: Any comments that can provide more clarity or information about the change request.

Change Log

A change log tracks the status of the change from submission to implementation or rejection. Table 18-2 shows an example of a change log.

TABLE 18-2 Change Log

ID	Submission Date	Description	Requestor	Status	Disposition

Requirements Traceability Matrix

Requirements traceability matrix: A grid that links requirements to deliverables and other project elements.

When a change request is submitted, the CCB will need to analyze the impact of that potential change. With large projects with hundreds of requirements it is very likely the CCB could miss the impact of the requested change on a requirement. One of the tools that can help you understand the impact of potential changes on a requirement is a requirements traceability matrix. You can create one to include whatever information you think is most relevant for tracking. Table 18-3 shows an example that traces requirements to business objectives and deliverables.

MANAGING CHANGE IN A HYBRID ENVIRONMENT

The change management plan, change request, and change log are primarily used for predictive deliverables. They can be adapted to fit a hybrid environment by determining when an adaptive change should have a change request and come to the CCB. The bureaucracy of the change management process needs to be balanced with the value it provides in controlling unwanted changes, protecting baselines, and integrating changes across deliverables.

TABLE 18-3 Requirements Traceability Matrix.

ID	Source	Requirement	Priority	Business Objective	Deliverable

Change for Predictive Deliverables

During the change analysis process, the CCB will assess the impact to scope, schedule, cost, and other variables. Sometimes there will be minimal change, and sometimes a change will require a new baseline for scope, schedule, or cost. These types of changes usually require the approval from the project sponsor.

> **Scope creep:** An increase of product or project scope without the necessary cost or schedule increase.

One of the behaviors that leads to problems on project is a request for a modification that doesn't go through the change control process. In this situation there is no accountability for the decision to make the change, and it can lead to cost and schedule overruns. This situation is called scope creep. Changing two wine racks to secure wine racks is a potential example of scope creep. If the cost difference is only $100 or so and there is no schedule impact, it probably isn't a big deal. If the cost doubles and there is a schedule delay, then it is a problem.

Change for Adaptive Deliverables

In a purely Agile environment, you can make changes and add features for as long as someone is willing to continue with the project. The potential downside of this is if you have an initial set of requirements and features for a product and you set an expected launch or release based on that, and then you continue to add features—then your launch or release date will be pushed out. Theoretically, this could continue until what you are really working with is *product management* rather than project management. There is nothing wrong with that, but it should be a deliberate choice rather than continuing to add features with no acknowledgment of the delayed finish date.

On a hybrid project, you can't just keep adding features, time, and cost because with a hybrid project there may be other deliverables that are waiting on the adaptive work. If the adaptive deliverable continues to change, this could have impacts on the rest of the project.

HELPFUL TOOLS

The best way to keep track of everything you need to manage on a hybrid project is to be organized. Logs and registers are a great way to help you track all things you need to stay on top of. Throughout this book we have described various logs and registers, such as:

- Assumption log (Chapter 4);
- Backlog (Chapter 6);
- Stakeholder register (Chapter 11);

TABLE 18-4 Decision Log

ID	Decision	Responsible party	Date	Comments

- Risk register (Chapter 14);
- Change log (this chapter).

There are three additional logs that can help you manage your project: A decision log, issue log, and impediment log.

Decision Log

If you've ever been in a situation where your team and you spend time working through alternatives, discussing risks and benefits, and finally reaching a decision, and then three months later, no one remembers what you decided, you can see the value in a decision log. A decision log is also valuable when you are meeting with your sponsor or the product owner. You will want to keep them apprised of decisions the team has made to make sure there isn't any relevant information you weren't aware of. Table 18-4 shows a sample decision log.

Issue Log

An issue is a current condition that may have an impact on a project. These are usually prioritized by how soon the issue is likely to occur, how soon you need a resolution, or how soon you need to take action. You can see a sample of an issue log that is used to track active issues in Table 18-5.

You can also add columns for responsible party and any other fields that are appropriate.

TABLE 18-5 Issue Log

ID	Issue	Priority	Impact	Due	Status	Resolution	Comments

TABLE 18-6 Impediment Log

ID	Date	Impediment	Priority	Status	Comments

Impediment Log

An impediment log is used for Agile deliverables. It is similar to an issue log. In fact, you can use an issue log to record impediments if you like. If you want to keep a separate log for impediments, Table 18-6 shows an example.

SUMMARY

In this chapter we discussed how to manage change for a hybrid project. We described the contents of a change management plan, a change request, and a change log. We described how a requirements traceability matrix can help you understand the implications of a change request on requirements and deliverables.

We described the differences in managing change for predictive and adaptive deliverables and the need to balance the management needs of the larger project with adaptive deliverables. We finished the chapter by looking at decision logs, issue logs, and impediment logs that can help you stay organized during the project.

Key Terms

change control board
change management
configuration management
requirements traceability matrix
scope creep

19

Metrics for Predictive Deliverables

As your project advances you will want to check the progress. For predictive deliverables you will focus on cost and schedule measures and provide forecasts for future work. There are many ways to measure progress. This chapter covers looking at variance from the critical path and how resource variances influence cost variances. Earned value management entails four key metrics and several calculations to provide performance information for schedule and cost status. You can also use these metrics to forecast cost information.

In this lesson you will learn common performance measures for predictive deliverables. You will learn to calculate variances and efficiency indexes. You will also learn ways to forecast future performance given the current status of the project.

PREDICTIVE MEASURES

Most predictive projects measure progress against baselines, for example, comparing planned schedule progress to actual progress, or budget estimates to actual costs. There is an expectation that

Threshold: A value that indicates action is required.

the project will have some cost and schedule variance, but how much is too much? Organizations usually set cost and schedule performance thresholds, but if the organization doesn't have them, then the team will want to establish thresholds so they can determine how the project is doing. Common thresholds are:

Acceptable: Performance variance within ±5% indicates the project is going as planned. Variances less than 5% are not anything to worry about. This is often indicated as green on a "stoplight" chart.

At risk: Performance variance between 5 and 10% indicates that performance may be deteriorating. The project team should investigate what is causing the variance and what actions can be taken to reduce the variance. This is often indicated as yellow or amber on a "stoplight" chart.

Troubled: A performance variance greater than 10% indicates the project is not performing as planned. The project team should take corrective action to address the variance. At this point the performance is considered troubled. This is often indicated as red on a "stoplight" chart.

If one objective is more important than others, that objective might have tighter threshold variances. For example, if meeting the schedule dates is critical, then the "at risk" threshold might be 3–5% and any variance over 5% would be considered troubled.

We'll start by reviewing schedule performance and then move on to cost. After reviewing some schedule and cost measures we'll look at a set of metrics you can use that combine schedule and cost performance. This system is called earned value management. We will not cover technical or performance measures because those are completely dependent on the type of project you are doing.

Schedule Measures

Reviewing tasks to see if they start on time and finish on time is one way of evaluating schedule performance. You can also look at the duration and see if they took longer or shorter than planned. This gives you some information about the status of the project. To make that information more relevant, look at the status of the critical path. Tasks that start or finish late but have a lot of float usually don't affect the delivery dates. However, a task that is on the critical path and is late, even by only a few days, can have a negative impact on your schedule.

You can evaluate the schedule by comparing start and finish dates in a tabular chart, as shown in Table 19-1, or with a Gantt chart, as shown in Figure 19-1.

Figure 19-1 lets you compare the planned and actual start dates and the planned and actual finish dates. The status date for this table is July 31, 2023. You can see there are three areas (shown as shaded) that are of concern:

- Hotel framing finished a week later than planned. This will cause the trades, finish work, and furniture and fixtures to start late because they are dependent on this task.

TABLE 19-1 Schedule Status Table

Task	Duration	% complete	Planned start	Planned finish	Actual start	Actual finish
Hotel	130 days	42%	Mon 4/17/23	Fri 11/10/23	Mon 4/17/23	NA
Structure	80 days	85%	Mon 4/17/23	Fri 7/28/23	Mon 4/17/23	N/A
Frame	9w	100%	Mon 4/17/23	Fri 6/16/23	Mon 4/17/23	Fri 6/23/23
Roof	2w	100%	Mon 6/19/23	Fri 6/30/23	Mon 6/26/23	Fri 7/7/23
Exterior	4w	40%	Mon 7/3/23	Fri 7/28/23	Mon 7/10/23	N/A
Trades	6 wks	0%	Mon 7/31/23	Fri 9/8/23	NA	NA
Finish work	6 wks	0%	Mon 9/11/23	Fri 10/20/23	NA	NA
Furniture and Fixtures	3 wks	0%	Mon 10/23/23	Fri 11/10/23	NA	NA
Restaurant	85 days	0%	Mon 6/19/23	Fri 10/27/23	NA	NA
Frame	3 wks	100%	Mon 6/19/23	Fri 7/7/23	NA	NA
Trades	2 wks	0%	Mon 9/18/23	Fri 9/29/23	NA	NA
Finish work	2 wks	0%	Mon 10/30/23	Fri 11/10/23	NA	NA
Furniture and Fixtures	2 wks	0%	Mon 11/13/23	Fri 11/24/23	NA	NA
Tasting Room	75 days	0%	Mon 7/17/23	Fri 12/1/23	NA	NA
Frame	2 wks	25%	Mon 7/17/23	Fri 7/28/23	NA	NA
Trades	1 wk	0%	Mon 10/2/23	Fri 10/6/23	NA	NA
Finish work	1 wk	0%	Mon 11/13/23	Fri 11/17/23	NA	NA
Furniture and Fixtures	1 wk	0%	Mon 11/27/23	Fri 12/1/23	NA	NA
Renovation	90 days	38%	Mon 4/10/23	Fri 8/11/23	Mon 4/10/23	NA
Blueprints	30 days	100%	Mon 4/10/23	Fri 5/19/23	Mon 4/10/23	Fri 5/19/23
Construction Review	0 days	100%	Fri 5/19/23	Fri 5/19/23	Fri 5/19/23	Fri 5/19/23
Production Facility	50 days	80%	Mon 5/22/23	Fri 7/28/23	Mon 5/22/23	NA
Structure	1 mon	100%	Mon 5/22/23	Fri 6/16/23	Mon 5/22/23	Fri 6/16/23
Trades	3 wks	100%	Mon 6/19/23	Fri 7/7/23	Mon 6/19/23	Fri 7/7/23
Finish work	1 wk	100%	Mon 7/10/23	Fri 7/14/23	Mon 7/10/23	Fri 7/14/23
Equipment	2 wks	0%	Mon 7/17/23	Fri 7/28/23	NA	NA

(Continued)

TABLE 19-1 Schedule Status Table *(Continued)*

Task	Duration	% complete	Planned start	Planned finish	Actual start	Actual finish
Storage Facility	40 days	63%	Mon 6/19/23	Fri 8/11/23	Mon 6/19/23	NA
Structure	1 mon	100%	Mon 6/19/23	Fri 7/14/23	Mon 6/19/23	Fri 7/14/23
Trades	2 wks	50%	Mon 7/17/23	Fri 7/28/23	Mon 7/17/23	NA
Finish work	1 wk	0%	Mon 7/31/23	Fri 8/4/23	NA	NA
Equipment	1 wk	0%	Mon 8/7/23	Fri 8/11/23	NA	NA
Construction Complete	0 days	0%	Fri 12/1/23	Fri 12/1/23	NA	NA

- The production facility equipment is supposed to be complete this week, but it has not started.
- The storage facility trades are supposed to be complete this week, but they are only 50% complete. This will have an impact on the storage facility finish work and equipment.

The Gantt chart in Figure 19-1 shows a visual representation of the status of the work.

Figure 19-1 makes it easier to see the slip for the tasks. This image is from Microsoft Project scheduling software. In the chart area, the dark rows on the bottom of task show the baseline duration of each task. The lighter lines on top of the baseline show the projected work times The top line is filled in according to the % complete. There is also a black vertical line that indicates the data date. The data date is the date for which the data is shown, in this case July 31, 2023. By looking at the chart you can see the following:

- For the Hotel Exterior, the line on top should be complete, but the progress indicator shows it is only 40% complete.
- The Production Facility Equipment shows no progress.
- The Storage Facility Trades baseline shows it should have finished in Mid-July, but the line on top indicates it hasn't finished yet.

Cost Measures

One of the most important and scrutinized measures on your project is cost. People want to know how much you have spent, how much more you are going to spend, and how much the total project will cost.

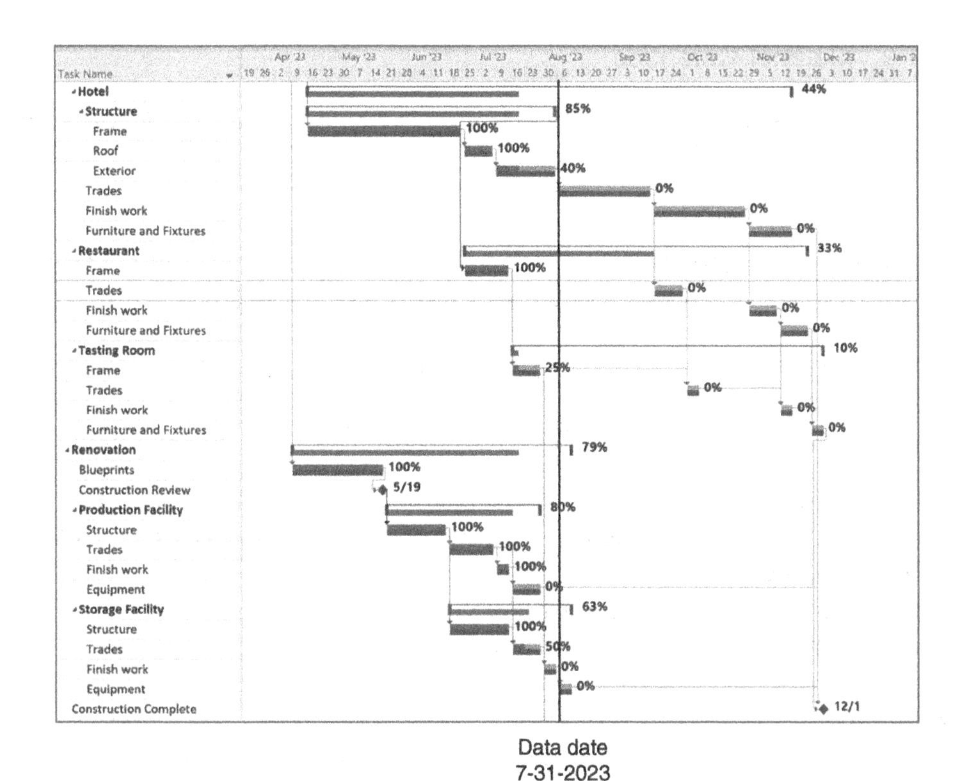

FIGURE 19-1 Schedule status Gantt chart.

Cost measures are essentially a function of comparing planned expenditures to actual expenditures. However, it can be useful to identify categories of cost variance, such material variances and labor variances. There are a few brief calculations you can perform to determine how much of a material variance is based on amount and how much is due to price. You can use similar calculations to determine how much of a labor cost variance is based on rate and how much on usage.

Material Variances

There are two sources of variance for materials. The first is using a different quantity of material, and the second is a difference in the price of material.

Using the Dionysus Winery tasting room as an example, the contractor estimated they would need 1500 2×4s with an estimated price of $2.05 each. Therefore, the estimated cost is $3,075. However, the invoice showed a cost of $3,597, which is a variance of –$522.

When asked about the difference the contractor stated there was more scrap and rework than planned. He says they used 1650 2×4s. When looking at the invoice you note it shows a cost of $2.18 for each 2×4 rather than $2.05.

TABLE 19-2 Material Variances

	Estimated Usage	Actual Usage	Estimated Price	Actual Price
2x4	1500	1650	$2.05	$2.18

To determine how much of the variance is due to usage and how much to price, you can set up a table like Table 19-2. Enter your estimated and actual usage and the estimated and actual price.

To calculate how much of the variance is due to usage, subtract the estimated usage from the actual usage, in this example that's 1650 – 1500, then multiply by the estimated price of $2.05. This equals $307.50. Thus $307.50 of the $522 variance is due to using more material.

To determine the impact of the price variance, subtract the actual price from the estimated price, $2.18 – $2.05, and then multiply by the actual usage of 1650. This indicates that $214.50 is due to price variance.

You can check your math by summing the variances to make sure they add up to the total variance. In this case, $307.50 + $214.50 = $522.

Labor Variances

If the cost variance is due to labor rather than materials, you can use the same calculations substituting hours for usage and rates for price. For example, the estimate for the interior designer for the hotel center was $7500 based on 100 hours and a rate of $75 per hour. She invoiced $8580. During a conversation she states there was a new requirement added and she needed 10 more hours to update the design drawings. In addition, her rate went up from the original quote that was obtained seven months ago. Based on this information you can set up a table like Table 19-3 to compare hours and rates.

Start by subtracting the estimated hours from actual hours (110 – 100), and multiply by the estimated rate of $75. This tells you $750 of the variance is due to more hours. Next subtract the estimated rate from the actual rate and multiply by the actual hours ($78 – $75) × 110. This shows that $330 was due to a rate difference. The sum of $750 + $330 = $1080.

TABLE 19-3 Labor Variances

	Estimated Hours	Actual Hours	Estimated Rate	Actual Rate
2x4	100	110	$75	$78

EARNED VALUE MANAGEMENT

The previous two sections looked at schedule status and cost status independent from each other. Often you can tell a better story if you combine information on scope, schedule, and cost, rather than looking at them in isolation. Earned value management (EVM) is a planning and tracking technique that allows you to combine scope, schedule, and cost information into a set of metrics so you can get an integrated view of your project. It is typically used on large projects with well-defined scope, such as construction or aerospace, but because the principles behind it are based on sound project management principles, it can be used on any predictive deliverables, even for small projects.

Earned value management (EVM): A method used to plan and track project performance.

We will describe three types of metrics, variances, indexes, and forecasts. We'll use the construction work for the Dionysus Winery project to talk through the concepts.

Planning for Earned Value

Earned value management starts with decomposing and organizing the scope with a WBS. The WBS for the Dionysus Winery was described in Chapter 6. It is shown again in Figure 19-2.

Next you schedule the work. The schedule was shown previously in Table 19-1 and Figure 19-1. The next step is to develop cost estimates for the work. Table 19-4 shows the cost estimates.

The cumulative total for all the work is $7,110,000. In earned value management, this is called the budget at completion, abbreviated as BAC. So far, these are the same steps you would take to plan any predictive work. What makes earned value management useful is taking the information from the scope, schedule, and cost estimates and combining it into an integrated baseline called a performance measurement baseline (PMB).

Budget at completion (BAC): The total cost for the work to be done.

To create a performance measurement baseline, you will work in a spreadsheet, such as Excel. Then simply follow these steps:

1. Enter the work in the left-most column.
2. Create a row for time, in this case March through December.

Performance measurement baseline (PMB): A baseline developed by integrating scope, schedule, and cost information.

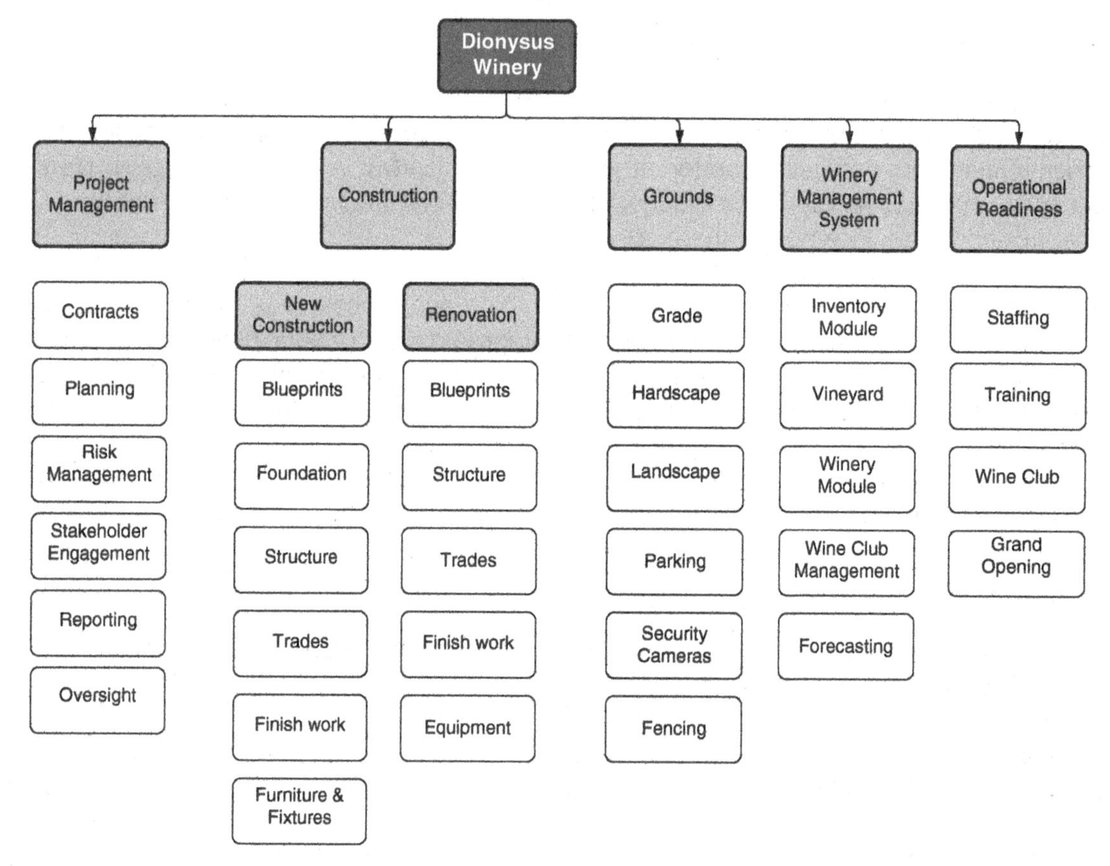

FIGURE 19-2 Dionysus Winery WBS.

3. Enter the value of the work planned for each work package in the appropriate month. For example, the exterior work for the hotel is scheduled for July 10 to August 4. It is 4 weeks in duration. About 75% of the work will be done in July and 25% in August. The value of the work is $1,200,000. Therefore we will enter $900,000 (75% of $1,200,000) in July and $300,000 in August.
4. Create a row after all the work packages and label it "Cost per Month." Sum the value of all the work planned for each month.
5. Create a row under the "Cost per Month" called "Cumulative Cost." Sum the previous month's work with the current month's work to get a cumulative cost to date for each month.

TABLE 19-4 Dionysus Winery Cost Estimates

Task	BAC
Hotel	**5,200,000**
Frame	800,000
Roof	250,000
Exterior	1,200,000
Trades	900,000
Finish work	850,000
Furniture and fixtures	1,200,000
Restaurant	**750,000**
Frame	175,000
Trades	165,000
Finish work	110,000
Furniture and fixtures	300,000
Tasting room	**200,000**
Frame	45,000
Trades	40,000
Finish work	40,000
Furniture and fixtures	75,000
Renovation	**962,000**
Blueprints	23,000
Construction review	2,000
Production facility	**495,000**
Structure	150,000
Trades	75,000
Finish work	50,000
Equipment	220,000
Storage facility	**440,000**
Structure	140,000
Trades	50,000
Finish work	30,000
Equipment	220,000
Total	**7,110,000**

Table 19-5 shows what the completed chart should look like.

TABLE 19-5 Performance Measurement Baseline Table

	April	May	June	July	August	September	October	November
Hotel								
Frame	200,000	300,000	300,000					
Roof			125,000	125,000				
Exterior				900,000	300,000			
Trades					450,000	450,000		
Finish work						350,000	500,000	
Furniture and fixtures								1,200,000
Restaurant								
Frame			100,000	75,000				
Trades						165,000		
Finish work								110,000
Furniture and fixtures								300,000
Tasting room								
Frame					45,000			
Trades							40,000	
Finish work								40,000
Furniture and fixtures								75,000
Renovation								
Blueprints	10,000	13,000						
Construction review		2,000						
Production facility								
Structure		75,000	75,000					
Trades			50,000	25,000				
Finish work				50,000				
Equipment				220,000				

TABLE 19-5 Performance Measurement Baseline Table *(Continued)*

	April	May	June	July	August	September	October	November
Storage facility								
Structure			70,000	70,000				
Trades				50,000				
Finish work					30,000			
Equipment					220,000			
Construction complete								
Cost per month	210,000	390,000	720,000	1,515,000	1,045,000	965,000	540,000	1,725,000
Cumulative cost	210,000	600,000	1,320,000	2,835,000	3,880,000	4,845,000	5,385,000	7,110,000

To make the table into a chart, like the one shown in Figure 19-3, follow these steps:

1. Copy the row with the months and the row with the cumulative cost per month.
2. Paste those two rows with the months on top and the cumulative cost per month on the bottom.
3. Insert a line chart.
4. You can add data labels to make it easier to read.

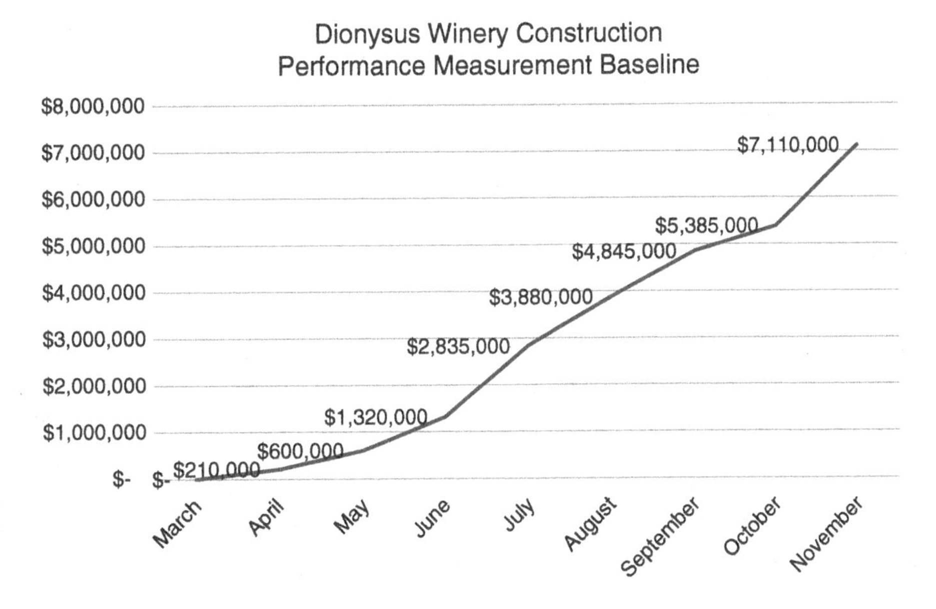

FIGURE 19-3 Performance measurement baseline chart.

Planned value (PV): The value of work expected to be accomplished.

This is a performance measurement baseline that you can use to track your project performance. You can see the planned value for the project for any given point of time. Planned value means the budget for the scheduled work. For example, the planned value for the hotel roof at the end of July is $250,000. The cumulative planned value at the end of August is $2,835,000.

A summary of the steps to create a performance measurement baseline are:

1. Organize your work with a WBS;
2. Schedule the work;
3. Estimate costs for the scheduled work;
4. Calculate the cumulative cost by time period;
5. Create a chart to show the performance measurement baseline.

Determining Earned Value and Actual Cost

Earned value (EV): The value of work accomplished.

Once you have your performance measurement baseline and you can see the planned value for each deliverable and at any point in time, you can start comparing the planned value to the value of the work accomplished. That is called the earned value.

Note: Earned value management is a planning and tracking management technique. Earned value is a measure of the work accomplished. Earned value management is sometimes known as earned value analysis or earned value measurement. Some people refer to earned value management as "earned value." It can be confusing, so listen to the context to understand if they are referring to a management system or the value of the work accomplished.

A simple way to determine the earned value for a deliverable is to multiply the value of the completed deliverable by the percent complete. For example, if the roof has a value of $250,000 and it is complete, then the earned value is $250,000. If the exterior work has a value of $1,200,000 and it is 50% complete, it has an earned value of $600,000.

Actual cost (AC): The funds spent to accomplish work.

Actual cost is determined by summing up all the costs associated with the work. This includes labor, material, licenses, travel, and any other expenses attributed to a deliverable. You usually determine the actual cost by looking at invoices.

Calculating Schedule and Cost Variances

With these three numbers, planned value, earned value, and actual cost, you can figure out schedule and cost variances. When calculating a variance, always start with the earned value and then subtract. When you are looking at a schedule variance, you subtract the planned value. The equation is EV – PV = SV. Here is some data for the Dionysus Winery at the end of August.

	PV	EV	AC
Frame	800,000	800,000	840,000
Roof	250,000	250,000	260,000
Exterior	1,200,000	900,000	985,000
Trades	450,000	0	0

Schedule variance (SV): The difference between the planned accomplishments and actual accomplishments. In earned value management, the difference between the earned value and the planned value (EV – PV).

Cost variance (CV): The difference between the actual costs and baseline costs. In earned value management, the difference between the earned value and the actual cost (EV – AC).

Using the schedule variance (SV) equation you can see there is a schedule variance of –300,000 for the exterior, and since the trades haven't started, there is a variance of –$450,000 for the trades. It can sound a little odd to say you are $300,000 behind schedule for the exterior, but this is because there is $300,000 less value produced than planned at that point in time.

When calculating the cost variance, you use a similar equation, but you subtract the actual cost from the earned value: EV – AC = CV. In the example above there are the following cost variances:

Frame: –40,000
Roof: –10,000
Exterior: –85,000

With the cost variances, notice you are comparing the work accomplished (EV) to the cost for the work (AC). You are not comparing what you spent (AC) to what you planned to spend (PV).

A negative variance indicates poor performance, while a positive variance indicates you are either ahead of schedule or under budget.

Calculating Schedule and Cost Indexes

> **Schedule performance index (SPI):** A measure of schedule performance efficiency calculated by the earned value divided by the planned value (EV/PV).
>
> **Cost performance index (CPI):** A measure of cost efficiency calculated by the earned value divided by the actual cost (EV/AC).

There is another way to consider the cost and schedule performance. You can use a ratio or index. Where a variance gives you a dollar number to indicate performance, an index tells you how efficient you are. Thus, a schedule performance index (SPI) indicates how efficient you are in meeting your schedule, and a cost performance index tells you how efficient you are in meeting your budget.

You use the same numbers you did for calculating the variances, but instead of subtracting, you divide. Thus, the equation for the schedule performance index (SPI) is EV/PV, and the equation for the cost performance index is EV/AC. Therefore, the SPI for the exterior is 900,000/1,200,000 = .75. You could say you are 75% efficient on the schedule for the exterior.

The CPI calculations for the winery are as follows:

Exterior: 900,000/1,200,000 = .75
Frame: 800,000/840,000 = .95
Roof: 250,000/260,000 = .96
Exterior: 900,000/985,000 = .91

For the exterior you are getting 75 cents of value for every dollar you spend. For the frame, you are getting 95 cents of value for every dollar spent. For the roof is 96 cents and for the exterior, 91 cents. A performance index less than 1.0 indicates poor performance, while a performance greater than 1.0 indicates you are either ahead of schedule or under budget.

> **Note:** The common practice is to calculate indexes to the 2nd decimal place.

The graph shown in Figure 19-4 is a common way of showing earned value data. You can see the earned value line is below both the planned value and the actual cost. Anytime the earned value is below something, it indicates unfavorable performance.

As mentioned previously, most organizations use thresholds of 0–5% variance as acceptable, 5–10% is at risk, and greater than 10% is a troubled project. The cumulative schedule performance for the hotel is $1,950,000/$2,700,000 = .72. Thus, this is a troubled project for schedule performance. The cumulative cost performance for the hotel is $1,950,000/$2,085,000 = .94. Therefore, the cost performance is at risk.

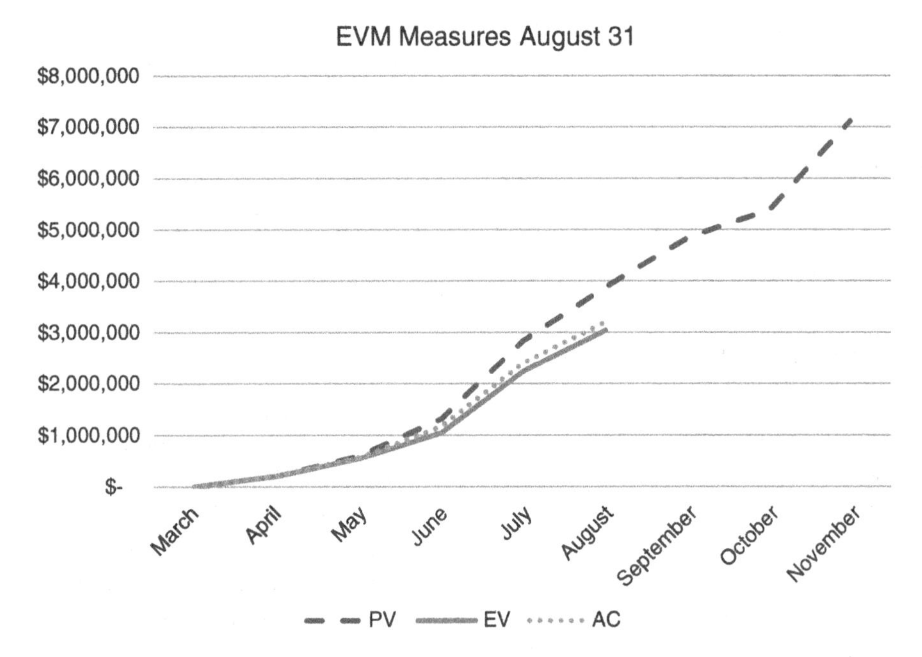

FIGURE 19-4 EVM Measures as of August 31.

There are a few important points to keep in mind when analyzing performance indexes:

- The CPI is one of the most reliable indicators of the health of a project. Usually about 20% of the way through your project the CPI stabilizes. In fact, if it changes, it usually gets worse because if you can't estimate the cost of work reliably for the work that is close in, you will likely do an even worse job for work that is farther out.
- The SPI indicates if work is going as planned. However, it does not take the critical path into consideration. You can have an SPI of less than 1.0 for a deliverable, but if there is float for that deliverable, you aren't really behind schedule. Therefore, you should always check to see if the schedule variance is on the critical path or a path with float.
- At the end of the project, your SPI will be 1.0 because all the planned work will have been accomplished; however, that doesn't mean your project was on time. Keep in mind that toward the end of the project, the SPI will trend toward 1.0. Always integrate an SPI with a critical path analysis to get a true picture of schedule status.

FORECASTS

Two of the most common questions you will hear are: "How much more do you expect to spend?" And "What will the final cost be?" You can use the information you gained with earned value management to develop forecasts that answer those questions. For the calculations for forecasting we will use the following cumulative information for the project as of August 31:

- Budget at completion = $7,110,000
- Earned value = $2,967,500
- Actual cost = $3,128,000
- Cost performance index = .95

Estimate to Complete

> **Estimate to complete (ETC):** The expected cost to finish the remaining work.

Calculating an estimate to complete (ETC) answers the question, "How much more do you expect to spend?" There are three ways you can answer this question. The first way is to take what you have learned from the project performance to date and develop an estimate for the funding you need to complete the project. For a large project, that could be very time consuming and perhaps somewhat subjective.

Another option is to subtract your earned value from your budget at completion; this gives you an estimate for the work remaining. For the Dionysus Winery construction, you would subtract $2,967,500 from $7,110,000 to arrive at an estimate to complete (ETC) of $4,142,500. That assumes that any variances you are currently experiencing would not carry forward. In other words, it assumes estimates for future work are accurate. This is not usually the case, so this may not be overly optimistic.

> **Note:** The equation for work remaining (WR) is BAC – EV. It is used in most forecasts.

A third possibility is to divide the work remaining by the current cost performance index, which assumes that future work will have the same level of cost performance as the past work. To calculate the ETC with this method, take the work remaining, in the case of the Dionysus Winery it is $4,142,500, and divide by the CPI of .95. This results in an estimate to complete of $4,360,526. The equation is (BAC – EV)/CPI. This is usually a more accurate estimate.

Estimate at Completion

> **Estimate at completion (EAC):** The expected total cost of the project.

There are many ways to answer the question, What will the final cost be? This is known as the estimate at completion (EAC). We will look at three ways to calculate the estimate at completion.

The first way (EAC math) assumes your original estimates were correct and any variances are a one-time event. You take the work remaining $4,142,500 and add the actual cost $3,128,000. This gives you an EAC of $7,270,500. The equation is BAC – EV + AC. This is usually not very realistic.

A more realistic way is called EAC CPI. You take the budget at completion (BAC) and divide it by the cost performance index (CPI). This assumes future performance will be at the same rate as current performance. For our project the EAC would be $7,110,000/.95 = $7,494,551.

If you are in a situation where you have both cost and schedule variances, the final cost usually reflects the schedule variance because you are either spending money for a longer period of time if you are late, or you are spending money to crash the schedule. This is referenced as EAC CPI × SPI. Many people think this is the most realistic way to calculate the EAC.

To calculate EAC CPI × SPI, start with the work remaining and divide it by (CPI × SPI). Then add the actual cost. The equation is

$$\frac{BAC - EV}{CPI\ X\ SPI} + AC$$

For the winery, the calculation is

$$\frac{\$4{,}142{,}500}{(.95 \times .75)} + 3{,}128{,}000 = \$8{,}942{,}035$$

For the Dionysus project there is a range of EACs from $7,270,500 to $8,942,035. It's a best practice to show a range of EACs and then indicate the one you think is most accurate, along with your reasoning.

> **Note:** You may get slightly different numbers if you use a spread sheet for calculations. Spreadsheets carry the decimals out further than two places. Don't worry about being precise, the intent is to show a range of possible outcomes that reflect the current state of the project.

Earned value management is a powerful technique to plan and track your project. That being said, it only tells you the state of your project; it does not tell you what caused the variances. You will have to investigate to find out what caused the current performance and what you can do about it.

SUMMARY

In this chapter we talked about measuring schedule and cost variances. We described common threshold measures for taking action, and we looked at a schedule that showed progress against a baseline.

We introduced earned value management along with a performance measurement baseline and four key measures: budget at completion, planned value, earned value, and actual cost. Then we described how you can use these measures to assess cost and schedule variance and cost and schedule performance. You can use the same measures to forecast how much more money you will need to finish the project (ETC) and to estimate the total cost of the project once it is complete (EAC).

Key Terms

actual cost (AC)
budget at completion (BAC)
cost performance index (CPI)
cost variance (CV)
earned value (EV)
earned value management (EVM)
estimate at completion (EAC)
estimate to complete (ETC)
performance measurement baseline (PMB)
planned value (PV)
schedule performance index (SPI)
schedule variance (SV)
threshold

20

Metrics for Adaptive Deliverables

Teams that work with adaptive deliverables focus on efficiency. The metrics they use track velocity, the time it takes to get a new feature developed, and estimates for the future rate of work.

In this chapter we'll look at how to use burn charts to estimate the amount of work a team can get done in an iteration, and we'll look at how to estimate a team's rate of work (velocity). Then we'll show how a cumulative flow diagram can identify bottlenecks in a process, show how long the team is working on new features, and indicate how long customers have to wait before their requests are fulfilled.

Agile projects often use measures to evaluate stakeholder satisfaction. We will look at how these measures can be used on any project and why they are definitely useful for hybrid projects.

ADAPTIVE MEASURES

Agile teams focus on increasing efficiency, throughput, and velocity. Noting that defects and rework take time to fix, they are also focusing on how to improve their work process to decrease defects. We'll describe four charts that are commonly used to plan and track progress for Agile deliverables, burndown charts, burnup charts, charts to estimate velocity, and a cumulative flow diagram.

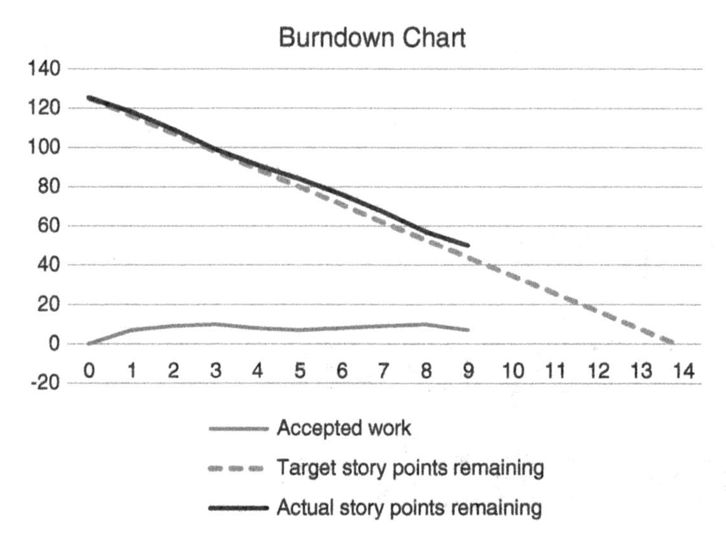

FIGURE 20-1 Burndown chart.

Burndown Charts

> **Burndown chart:** A graph that maps the actual work completed to the target work to be completed.

A burndown chart shows the work and an estimated completion rate. The team then tracks the actual work completed against the target completion. The burndown chart in Figure 20-1 shows information for a team that is using story points to estimate work. Their iterations (sprints) are one week in duration. They have 125 story points they need to accomplish. They estimate that they can accomplish nine story points per iteration.

The dashed line represents the target story points; you can see how it goes down by nine points each iteration. That is the target the team is working toward. The line at the bottom shows the work that has been accepted by the product owner each iteration. The solid line shows the actual work completed. This is the difference between the 125 points and the work that has been accepted so far. When the solid line is above the dashed line, the team is behind schedule.

This type of chart can be created on a flip chart, and the progress is charted at the end of each iteration. This is known as a low-tech, high-touch tool. In other words, it is just paper and marker (low tech) that is kept up to date by manually updating information (high touch).

You can also create this type of tool in a spreadsheet, such as Excel, and keep it on a team collaboration site if the team works remotely. To create this type of tool in a spreadsheet, follow these steps.

Create the following columns:

1. **Iteration:** If you aren't working with iterations, you can use time, such as days, weeks, or months. This is shown along the bottom of Figure 20-1.
2. **Target:** This is the amount of work you expect to accomplish in each time period. This is also known as target velocity. This is not shown in Figure 20-1; it was hidden to simplify the chart.
3. **Accepted work:** This is the point value of the work the product owner accepts as complete. This is shown as the line along the bottom of Figure 20-1.
4. **Target story points remaining:** To get this number start with the total number of story points, in this case 125, in the first row (the first row is iteration 0). Then subtract the target work for the first iteration to get the estimated story points remaining at the end of the first iteration. This is the dotted line in Figure 20-1.
5. **Actual story points remaining:** This is the total number of story points (entered in iteration 0) minus the accepted work. This is the decreasing solid line in Figure 20-1.

Table 20-1 shows the what the table looks like before inserting a line chart.

TABLE 20-1 Burndown Chart Table

Iteration	Target	Accepted work	Est. story points remaining	Act. story points remaining
0	0	0	125	125
1	9	7	116	118
2	9	9	107	109
3	9	10	98	99
4	9	8	89	91
5	9	7	80	84
6	9	8	71	76
7	9	9	62	67
8	9	10	53	57
9	9	7	44	50
10	9		35	
11	9		26	
12	9		17	
13	9		8	
14	9		-1	

One of the problems with a burndown chart is that if someone adds a new feature, it looks like the team's velocity slowed down, because there are more estimating remaining story points. To accommodate that, it might be better to use a burnup chart instead.

Burnup Charts

Burnup chart: A graph that shows the total work along the top and a line that shows the work completed.

As you can see in Table 20-2, a burnup chart mostly uses the same information as a burndown chart, but instead of tracking the work remaining, it tracks the work accomplished. Here is the table for a burnup chart; notice that the column for the total story points stays the same, but you can see at iteration 6, 10 story points worth of work was added, and at iteration 8, another 5 story points were added.

TABLE 20-2 Burnup Chart Table

Iteration	Total story points	Target work	Target work accepted	Accepted work	Actual work accepted
0	125	0	0	0	0
1	125	9	9	7	7
2	125	9	18	9	16
3	125	9	27	10	26
4	125	9	36	8	34
5	125	9	45	7	41
6	135	9	54	8	49
7	135	9	63	9	58
8	140	9	72	10	68
9	140	9	81		
10	140	9	90		
11	140	9	99		
12	140	9	108		
13	140	9	117		
14	140	9	126		
15	140	9	135		
16	140	9	144		

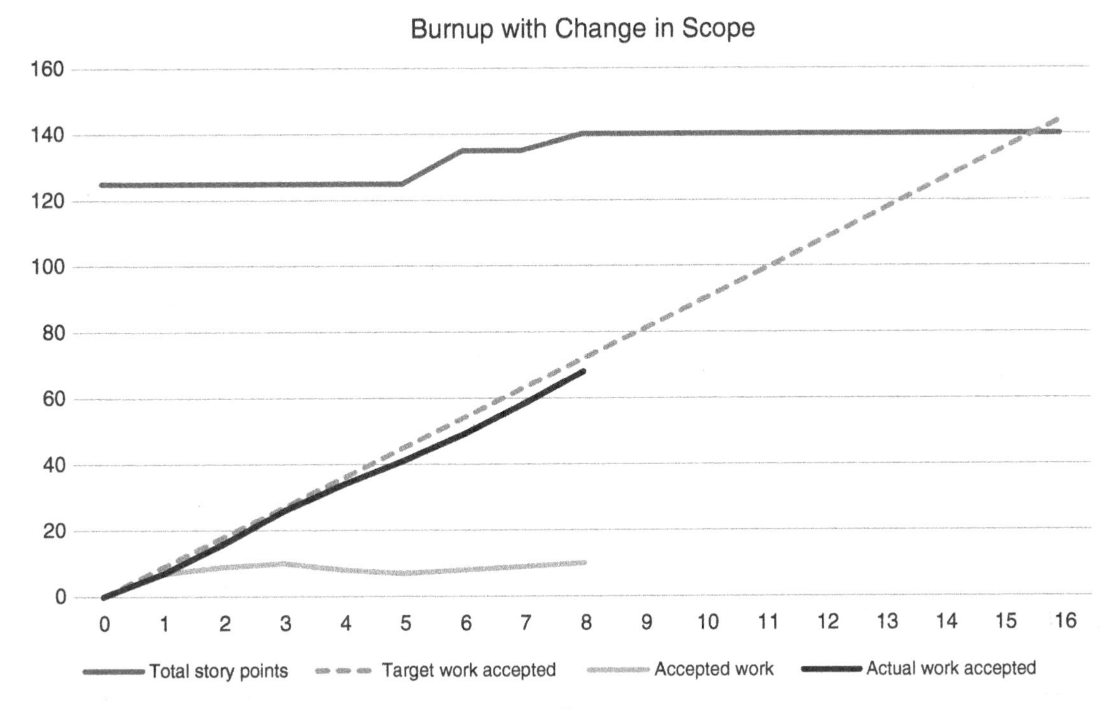

FIGURE 20-2 Burnup chart with change in scope.

Figure 20-2 shows how the chart looks. You can see the top line that represents the total story points line goes up in conjunction with the total story points column in Table 20-2.

Burnup charts provide a quick answer to the "When will it be done?" question by looking at the current rate of work compared to the target rate and forecasting the completion date given the team's velocity and the remaining work.

Estimating Velocity

> **Velocity:** The rate at which a team accomplishes work in a time period.

Velocity is the team's rate of work in a fixed time period, such as an iteration. It is used in planning and estimating as you saw with the burndown and burnup charts. It usually takes a few iterations for velocity to stabilize while the team gets familiar with the project, with each other, and figures out the best way to accomplish work. Once the team has worked together for a few iterations, it is relatively easy to estimate velocity for future sprints, and thus this is a great tool for forecasting how much work can be done in each iteration and the completion dates for releases.

Figure 20-3 shows a chart with rather uneven velocity in the first four iterations. Starting in iteration 5 they started tracking their average velocity, shown as the line. This shows that their velocity has averaged out to about 9 story points per iteration.

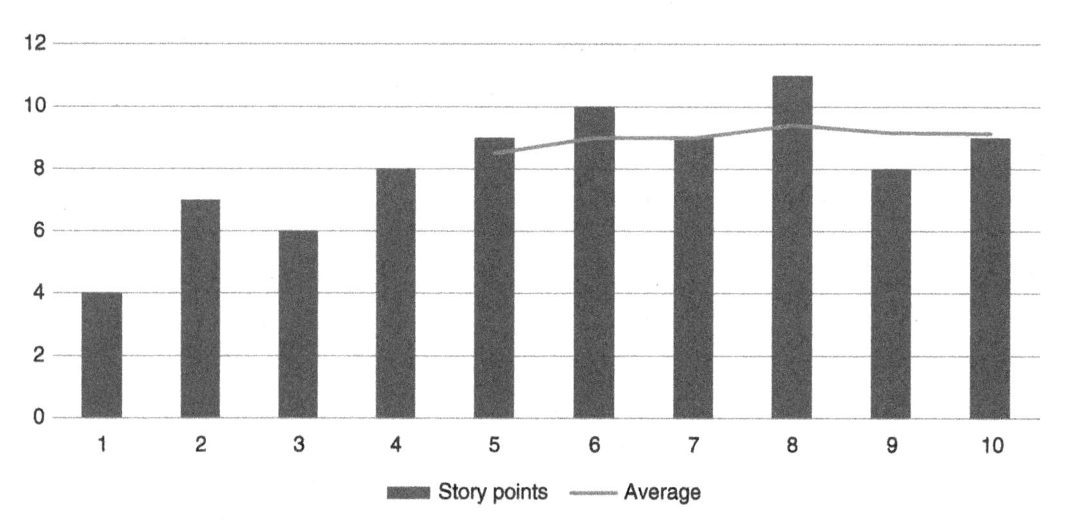

FIGURE 20-3 Velocity Chart.

CUMULATIVE FLOW DIAGRAMS

One of the principles for Agile teams is "Deliver working software frequently, from a couple of weeks to a couple of months, with a preference to the shorter time scale." To align with this principle, teams focus on increasing amount of work they can get done in a period of time, also known as throughput. They do this by limiting the work in progress so as not to create bottlenecks in the flow of work, by improving their work processes to decrease defects and increase the work done, and by reflecting on how to become more effective.

Cumulative flow diagram: A chart that shows the flow of work from backlog, through different stages of development, to completion.

Lead time: The time it takes to get a piece of work through the entire development process.

Cycle time: The time it takes to get a piece of work through a part of the development process.

A cumulative flow diagram helps visualize the flow of work. It shows the work in progress, the time it takes a task to get through the entire process from request to deployment (known as lead time), and the time work spends in a part of the process, such as development or testing (known as cycle time).

Figure 20-4 shows a cumulative flow diagram for building the wine club website. The time is across the horizontal axis and the story points are on the vertical axis.

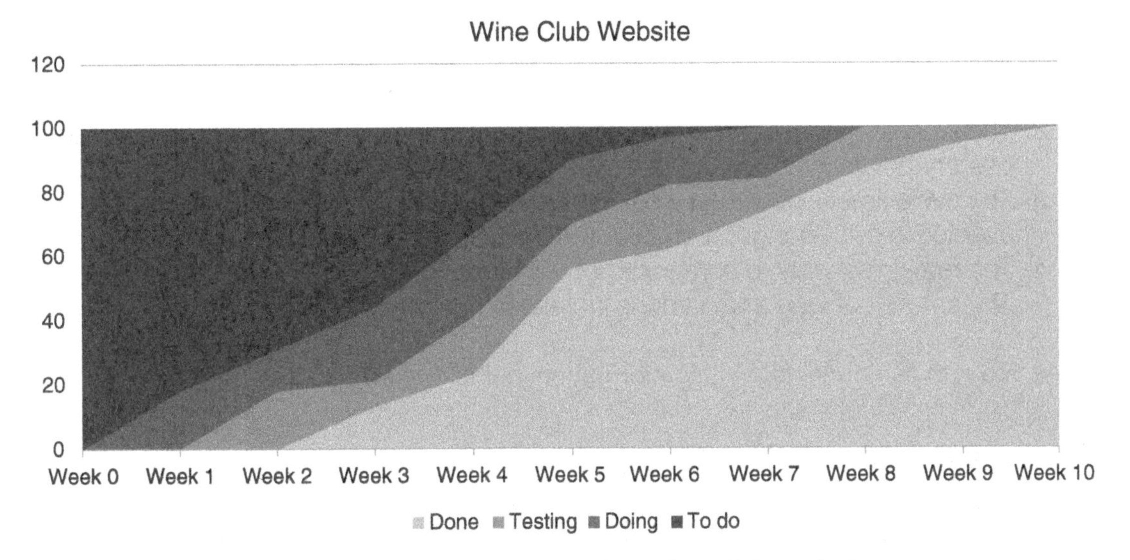

FIGURE 20-4 Cumulative flow diagram for the wine club website.

You can see the amount of work waiting to start is represented as the to-do work. Some work is in the "doing" section, some in the "testing" section, and some is done. To maintain a smooth flow of work, you want the work in the doing and testing sections to remain consistent. You can see around week 3 it looks like there is a bit of a bottleneck in the testing process. This is due to more work being started than there was work being tested. Figure 20-5 shows the same diagram with annotations to explain how to interpret the chart.

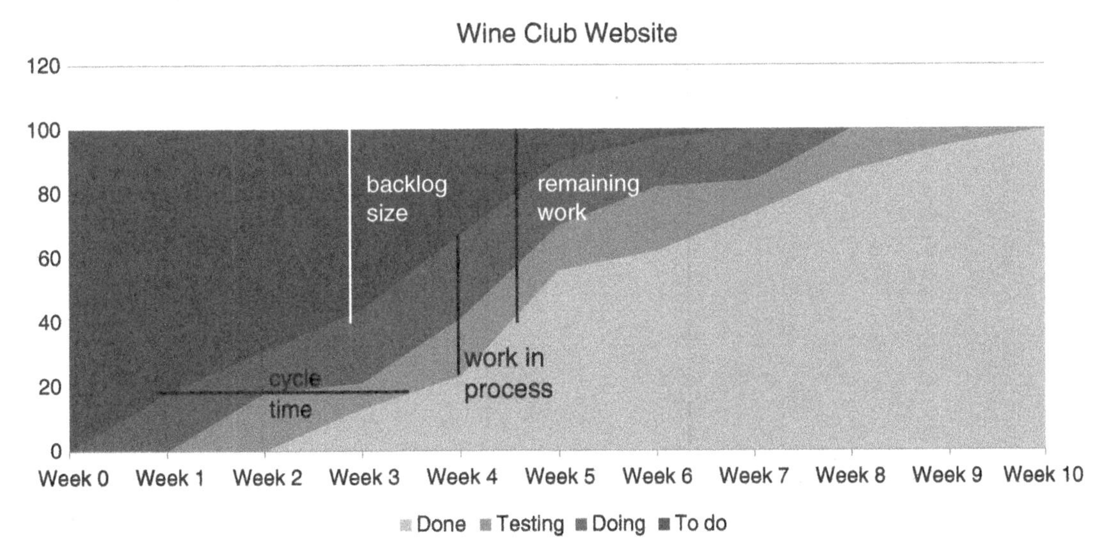

FIGURE 20-5 Cumulative flow diagram with annotations.

Note the following:

1. The backlog size is diminishing over time.
2. The work in process (WIP) at any given point in time is composed of the number of story points in doing and testing.
3. The cycle time is how long work spends in the doing and testing processes. The average is a little longer than two weeks in this example.
4. The remaining work is composed of the to-do, doing, and testing processes.
5. The amount of work that is done increases over time.

As you can see, there is a lot of information in this one chart.

Creating a Cumulative Flow Diagram

A cumulative flow diagram is useful but a little tricky to create. I'll show you how I created the chart for creating a website for the wine club members for the Dionysus Winery.

You start by get the data you will need for the chart from the task board. This example assumes you are using story points, but you can use duration estimates as well. Figure 20-6 shows the story points for 20 user stories that make up the work necessary to create a wine club member website.

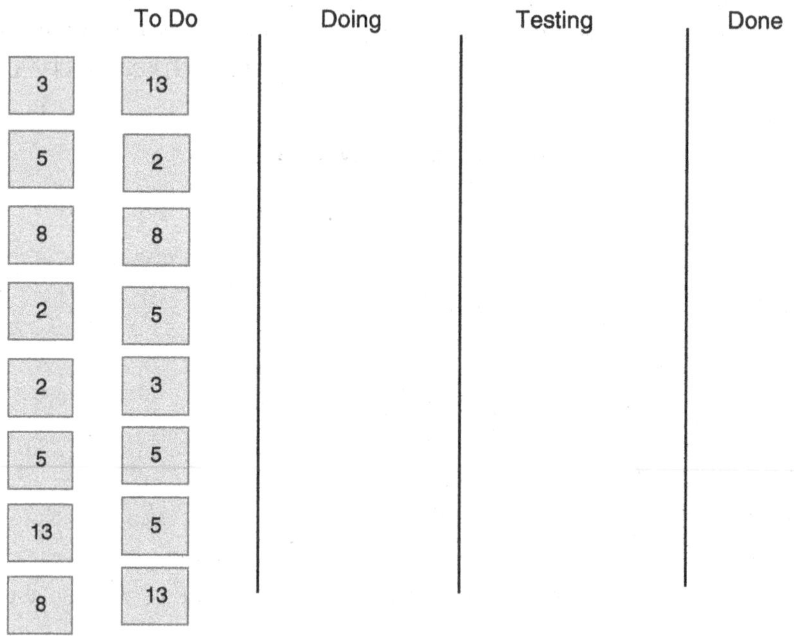

To Do
Doing
Testing
Done
3
13
5
2
8
8
2
5
2
3
5
5
13
5
8
13

FIGURE 20-6 Initial Task Board.

This shows there is 100 points worth of work to create the website. To create the cumulative flow diagram, you start with creating a table to enter in data.

1. First, create five columns. The first column will represent time. Enter in the time periods you are using. For this example, we assume each iteration is one week. Make sure to start with week 0 or iteration 0. This allows you to show the total number of points. Also make sure to enter words (week 0, sprint 0, etc.) so Excel doesn't think these are numbers it needs to track.
2. The next columns will reflect the columns for the task board, in this case, to do, doing, testing, and done.
3. Enter the total amount of work that needs to be done under week 0, to do. At this point the table should look like Table 20-3.
4. Throughout the project you will track the work that moves from the backlog (to do) into the doing column, then testing, and then done. Table 20-4 shows the data from the task board through week 4 and Figure 20-7 shows what the task board would look like at week 4.
5. To create the chart, select all the rows with data in them. Then go to the "Insert" menu and select a "Stacked Area" chart. In Excel you will find this under the "Line" group of charts. At first the chart looks like Figure 20-8. This is not particularly useful.

TABLE 20-3 Initial Cumulative Flow Diagram Table

Time	To do	Doing	Testing	Done
Week 0	100			
Week 1				
Week 2				
Week 3				
Week 4				

TABLE 20-4 Week 4 Cumulative Flow Diagram Table

Time	To do	Doing	Testing	Done
Week 0	100	0	0	0
Week 1	82	18	0	0
Week 2	69	13	18	0
Week 3	46	23	13	18
Week 4	33	26	18	23

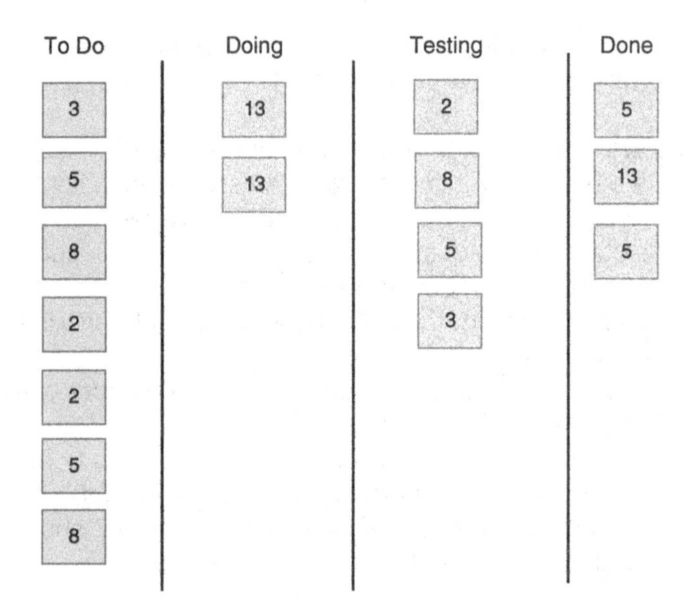

FIGURE 20-7 Task board at week 4.

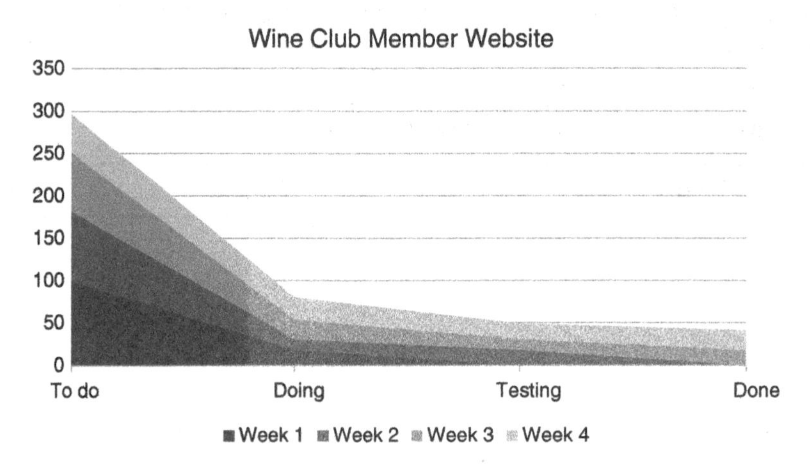

FIGURE 20-8 Initial Wine Club Member Website Cumulative Flow Diagram Week 4.

6. To format the chart to provide useful information right-click on the chart and choose "Select Data." When you do that, a "Select Data Source" box comes up. First, switch the row and columns. This will send the weeks to the bottom.
7. Next, reverse the order of the data. You can do that by selecting the legend entry and then clicking the up or down arrows to rearrange the data. Now the "to do" category is on bottom, then "doing," "testing," and "done" at the top. Once you have done that your diagram should look like Figure 20-9.

This cumulative flow diagram gives a much better picture of the team's workflow.

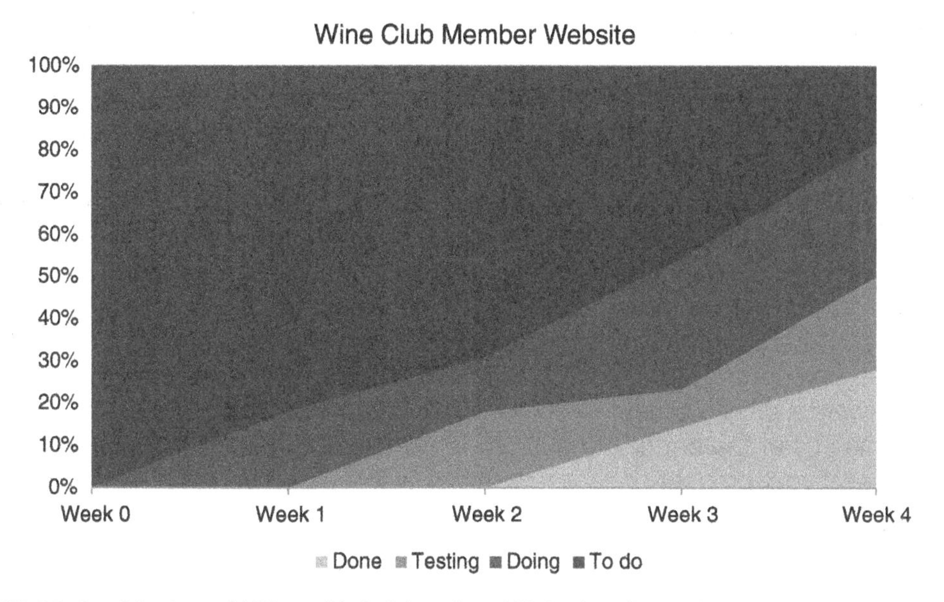

FIGURE 20-9 Updated Wine Club Member Website Cumulative Flow Diagram Week 4.

STAKEHOLDER MEASURES

A lot of project measures deal with information that is easy to quantify, like schedule and cost variances. But there are some really important metrics that have to do with your team and your stakeholders. We'll look at two of these measures, a Net Promoter Score® and a mood chart.

Net Promoter Score®

Net Promoter Score®: A measure of stakeholder loyalty and satisfaction.

A Net Promoter Score® measures customer satisfaction, but more importantly it measures how likely a customer is to refer someone else to use your product. Many hybrid projects have a digital component. Sometimes that component is for internal customers, and sometimes it is for consumers. The Net Promoter Score® is determined by asking stakeholders one simple question, "On a scale of 1–10, how likely are you to recommend this (product/service/company) to someone else?" The responses are usually categorized as:

Promoters: These people indicated a score of 9–10. This indicates an enthusiastic supporter who will share their positive experience with others.

Passives: These people scored the product or service as 7–8, indicating they are satisfied customers but not necessarily enthusiastic customers. Given a better option, these customers may choose a competitor.

Detractors: A score of 1–6 indicates dissatisfied customers. These people may spread negative opinions about their experience.

To develop the Net Promoter Score®, you determine the percent of people who are promoters and the percent who are detractors. Then subtract the percent of detractors from the percent of promoters. Theoretically, this leads to a range of 100% if every respondent were a promoter and –100% if every respondent were a detractor. Obviously, the higher the score, the better the outcome. It is useful to track this measurement periodically to see if your product or service is improving or deteriorating.

Mood Chart

Mood chart: A visual representation of team member moods or attitudes.

Your team is your most valuable resource, so you want to make sure they are happy, or at least satisfied. One way to assess team attitude is to use a mood chart. A mood chart lets people show you how they feel, and it is low tech and can be fun as well. Most people use emojis for a mood chart, but you can use colors or numbers as well.

You simply ask your team members to draw an emoji that reflects their mood. You can ask them to fill it in daily, weekly, or whenever you think it is important to get a sense of your team's attitude. Figure 20-10 shows a mood chart using emojis.

Team Member	Monday	Tuesday	Wednesday	Thursday	Friday
Issa					
Emily					
Mahmoud					

FIGURE 20-10 Mood chart.

Another way to check in with your team is to send out a quick survey asking them to use a scale of 1–5 to indicate how much they agree with statements such as:

- I feel appreciated;
- I think my work makes a difference;
- I enjoy working with my team.

Collecting information on team member and customer satisfaction is usually done through questionnaires, surveys, interviews, and conversations. While this type of data is more subjective than cost and schedule variances, it provides excellent information to evaluate how well the team is doing.

SUMMARY

In this chapter we looked at four charts to track progress for adaptive work. We looked at burndown and burnup charts, which compare target work to actual work. We used a bar chart over time to estimate the team's velocity. We also looked at a cumulative flow diagram to look at the flow of work through the categories on a task board. We can use this to identify bottlenecks, lead time, and cycle time. We wrapped up this chapter by looking at two stakeholder measures, a Net Promoter Score® and a mood chart.

Key Terms

burndown chart
burnup chart
cumulative flow diagram
cycle time
lead time
mood chart
Net Promoter Score®
velocity

21

Reporting for Hybrid Projects

In the two previous chapters we looked at metrics and measures for predictive deliverables and adaptive deliverables. Once you have that information, you need to display it in a format that makes it easy for your stakeholders to understand. Having easy-to-understand reports helps you determine the appropriate next steps. For example, do you need to take corrective action? Do you need to intervene to keep the project on track? Or is everything going fine?

In this chapter we will look at reporting. We will start by introducing three narrative reports, and then we will discuss visual reporting. Visual reporting includes project dashboards and information radiators. Then we will discuss some tips for creating hybrid dashboards.

REPORTING

The purpose of reporting is to provide information so you can take appropriate action. There are many different types of reports you can use; we'll categorize them into three groups: narrative reports, dashboards, and big visible charts. Narrative reports are text based, while dashboards and big visible charts provide a visual display of information. The visual displays provide a synopsis of information in an easy to digest and remember format. We'll start by looking at narrative-based reports, and then we'll discuss visual-based reports.

Narrative Reports

Narrative reports provide a description of the project's state. They typically describe progress made since the previous reporting period. Narrative progress reports are useful if you want to provide explanations, context, and detailed information on the status of the project. Where possible, they should be no longer than one page. Common narrative reports include status reports, variance reports, and earned value analysis reports.

Status Reports

A status report describes what has happened since the previous reporting period, any variances, and necessary actions to address variances. They may also identify new risks or issues since the previous report. Table 21-1 describes the content typically found in a status report.

This type of report is good to provide a general overview of the project.

Variance Reports

A variance report identifies planned and actual results for schedule, cost, and quality. It also describes the root cause and planned response. It is similar to a status report, but the focus is on identifying and responding to variances. A typical variance report contains the following information:

Schedule variance

- Planned results;
- Actual results;
- Variance;
- Root cause;
- Planned response,

Cost variance

- Planned results;
- Actual results;
- Variance;
- Root cause;
- Planned response.

Quality variance

- Planned results;
- Actual results;
- Variance;
- Root cause;
- Planned response.

TABLE 21-1 Status Report

Scope Information	
Accomplishments for this reporting period	All work and accomplishments completed in the reporting period
Changes to scope	Scope additions, deletions, or changes in the reporting period
Quality information	Defects or quality issues encountered in the reporting period
Schedule Information	
Accomplishments planned but not completed	All work and accomplishments scheduled for the reporting period but not completed
Root cause of variances	The cause of variance for work not accomplished as scheduled in the reporting period
Impacts to upcoming milestones or project due date	Impacts to milestones or the project schedule for work that was not accomplished as scheduled; work that is behind on the critical path
Planned corrective or preventive action	Actions needed to make up schedule variances or prevent future schedule variances
Cost Information	
Funds spent this reporting period	Documentation of funds spent in the reporting period.
Root cause of variance	The cause of variance for expenditures over or under plan; it includes information on labor variances and material variances and describes if the variance is due to basis of estimates or estimating assumptions
Impact to overall budget or contingency funds	Impact to the project budget and whether contingency funds must be expended
Planned corrective or preventive action	Actions needed to recover cost variances or prevent future cost variances
Look Ahead	
Accomplishments for next reporting period	Work and accomplishments scheduled for next period
Costs for next reporting period	Funds planned to be expended next period
Risks and Issues	
New risks identified	New risks identified this period
Issues	New issues from this period.
Comments	Comments that add relevance to the report

Earned Value Analysis Report

If you are employing earned value management (EVM), as described in Chapter 19, you can use an earned value analysis report. This type of report looks at metrics for the current period and the cumulative results. It usually compares the current cumulative results to the past period cumulative results so the reader can see whether trends are improving or deteriorating. Table 21-2 shows a sample earned value analysis report.

The percent planned, percent earned, and percent spent compares the current PV, EV, and AC to the BAC. For example, if you have a BAC of $1,000,000, PV of $450,000, EV of 420,000, and AC of $460,000, you will see these results:

Percent planned	45%
Percent earned	42%
Percent spent	46%

This shows that you have spent 46% of the budget to accomplish 42% of the work. Not a good state to be in. It also shows you should be 45% of the way through the project, but you are only 42% of the way through.

TABLE 21-2 Earned Value Analysis Report

Budget at completion (BAC):	________	Overall status:	________
	Current period	Current period cumulative	Past period cumulative
Planned value (PV)			
Earned value (EV)			
Actual cost (AC)			
Schedule variance (SV)			
Cost variance (CV)			
Schedule performance index (SPI)			
Cost performance index (CPI)			
Root cause of schedule variance			
Schedule impact			
Rood cause of cost variance			
Budget impact			
Percent planned			
Percent earned			
Percent spent			
Estimates at completion (EAC)			
EAC w/CPI (BAC/CPI)			
EAC w/ CPI × SPI			
[AC + ((BAC – EV)/ (CPI × SPI))]			
Selected EAC, justification, and explanation:			

The benefits of progress reports are that they provide good information on the project progress, and they allow you to explain variances and planned actions to correct performance. The downside is that they aren't very compelling, and they can take some time to read through.

VISUAL REPORTS

Visual reports are the equivalent of a picture being worth a thousand words. The intent of visual reports is to communicate a lot of information with few words. They provide a quick and easy way to absorb project information. Dashboards provide a variety of graphics in an electronic format, and information radiators provide a low-tech, high-touch way to see the status of a team's work.

Dashboards

A project dashboard provides an at-a-glance view of key performance indicators (KPIs), such as schedule, budget, and resource status. The purpose of a dashboard is to display critical information on a single screen. Dashboards should provide information clearly and with few distractions. The user should be able to quickly review and assimilate the information on the dashboard. Figure 21-1 shows a generic sample of some of the charts you

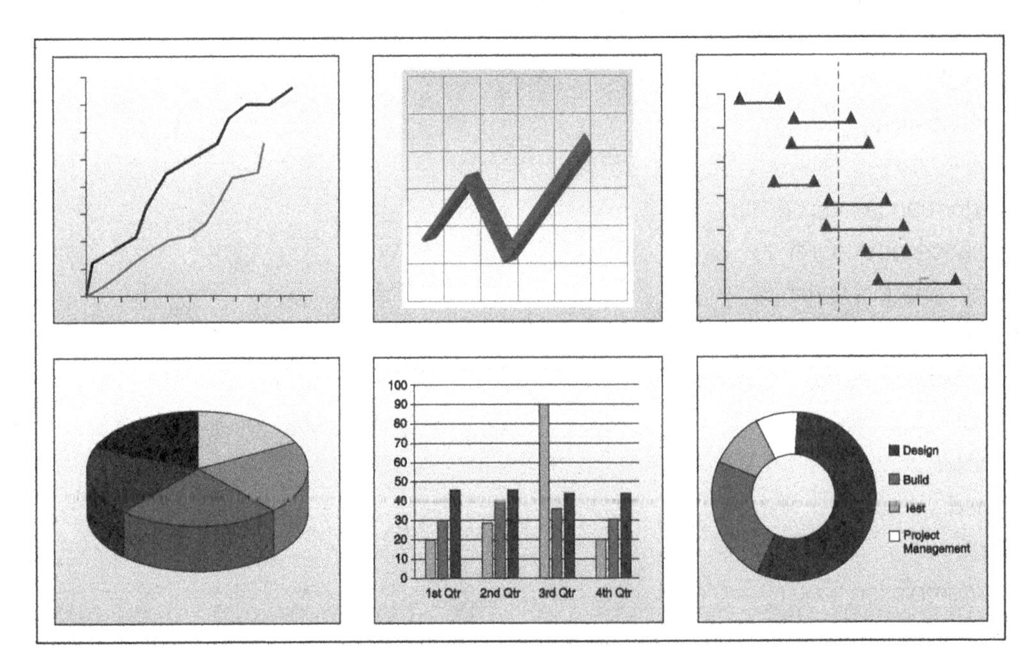

FIGURE 21-1 Sample dashboard.

Source: **Harold Kerzner 2013 / John Wiley & Sons.**

PROJECT OVERVIEW

MON 1/2/23 - FRI 1/12/24

% COMPLETE
Status for all top-level tasks. To see the status for subtasks, click on the chart and update the outline level in the Field List.

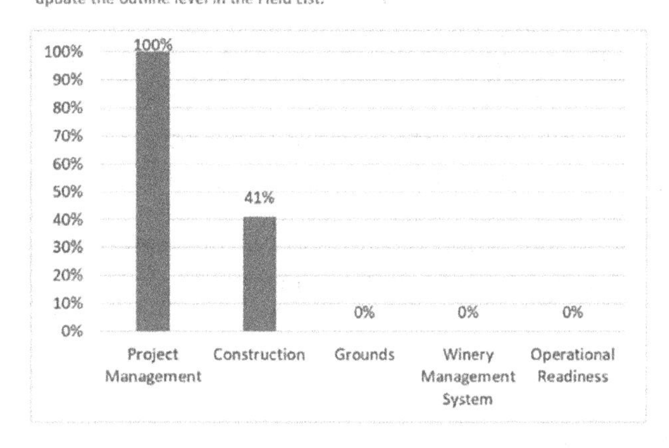

MILESTONES DUE
Milestones that are coming soon.

Name	Finish
Construction Complete	Fri 11/3/23
Grounds complete	Fri 9/22/23
System Readiness Review	Fri 11/17/23

LATE TASKS
Tasks that are past due.

Name	Start	Finish	Duration	% Complete	Resource Names
Structure	Mon 5/22/23	Fri 6/16/23	1 mon	75%	
Trades	Mon 6/19/23	Fri 7/7/23	3 wks	0%	
Structure	Mon 6/19/23	Fri 7/14/23	1 mon	0%	
Start up	Mon 6/12/23	Fri 6/23/23	2 wks	0%	
Release 1	Mon 6/26/23	Fri 8/4/23	6 wks	0%	

FIGURE 21-2 Schedule dashboard.

might find on a dashboard. Figure 21-2 shows a dashboard that focuses on schedule performance.

Creating a dashboard is a discipline unto itself. They are usually built with a spreadsheet or database to collect data and a graphical user interface to display it. There is specialized software such as Tableau or Power BI that specialize in turning data into dashboards. Rather than describing how to build dashboard we will discuss the types of information you can display on a dashboard.

Considerations for Creating a Dashboard

Before gathering all possible information to put on a dashboard, keep these four guidelines in mind:

1. **User needs:** Find out what your user needs are. You may need different dashboards for a steering committee than for team leads.
2. **The right metrics:** Focus on getting the right metrics—not all information is useful for a dashboard.
3. **Accurate information:** Make sure you have high-quality, accurate information. A dashboard with beautiful graphics is useless if the information is not up to date and accurate.
4. **Less is more:** The purpose of a dashboard is to provide an at-a-glance reference for project status. If you have too much information, or if the presentation is cluttered, people won't be able to absorb the information and it defeats the purpose of a dashboard.

Types of Charts

There are many different visual images you can use for a dashboard. We will review several of them in the next few pages.

Traffic Light

TABLE 21-3 Stoplight Chart

Grade	●
Hardscape	●
Landscape	●
Parking	●
Security cameras	●
Fencing	●

A traffic light or stoplight chart, as shown in Table 21-3, shows status as red, yellow, or green. These are sometimes also known as RAG charts, for red, amber, green.

Red: Typically red means there is a problem that will affect meeting a scope, quality, schedule, cost, or other objective. It could be the result of a risk that has materialized or a new issue that has arisen. A red indicator may necessitate getting the sponsor or customer involved.

Yellow/amber: This indicates there is a potential problem. With yellow status the project manager is looking for the root cause and identifying preventive actions so the performance does not deteriorate to red status.

Green: Green indicates everything is going according to plan.

These charts provide a simple top-level look at the status for various components. They also indicate if performance is approaching a threshold.

Line Charts

Line charts show changes over time. They can alert you to performance trends. Figure 21-3 shows a line chart with cost variances over time for three work packages.

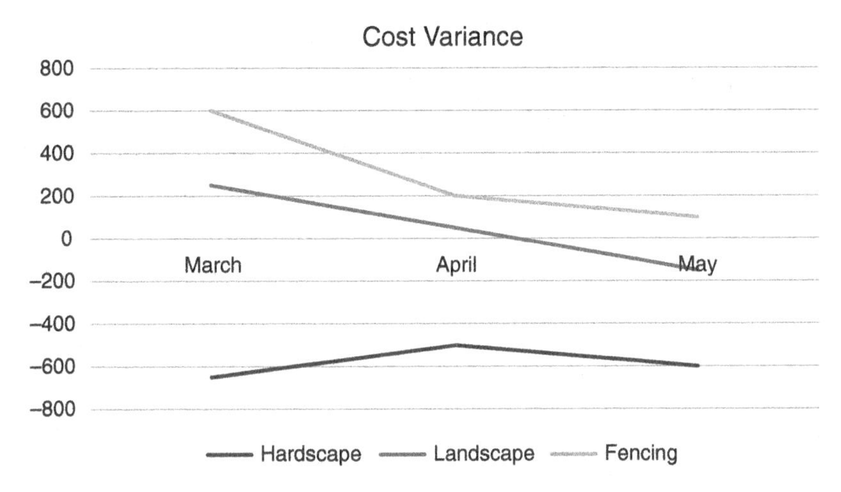

FIGURE 21-3 Line chart.

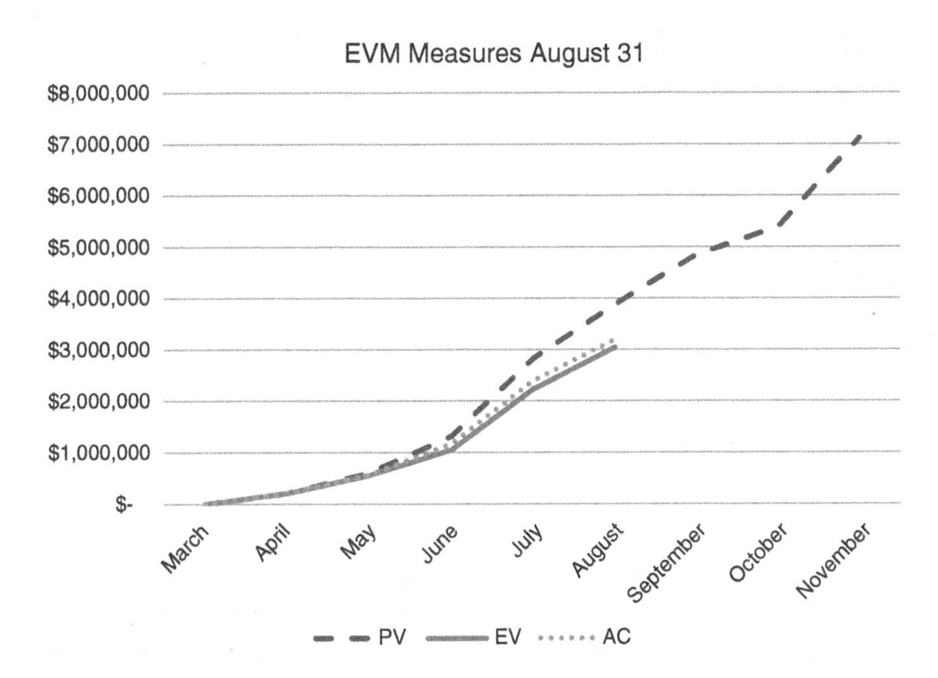

FIGURE 21-4 Earned value management line chart.

Burnup and burndown charts are examples of line charts. Figures 20-1 and 20-2 in the previous chapter show a burnup and burndown chart. As you can see in Figure 21-4, earned value management often uses a line chart as well.

Area Charts

An area chart is similar to a line chart, but the area is colored or shaded. The cumulative flow diagram shown in Figure 20-9 is an area chart. It is called a 100% stacked area chart because the different categories are stacked, and the chart shows percentages. If you use a regular stacked area chart, it will use values rather than percentages, as shown in Figure 21-5.

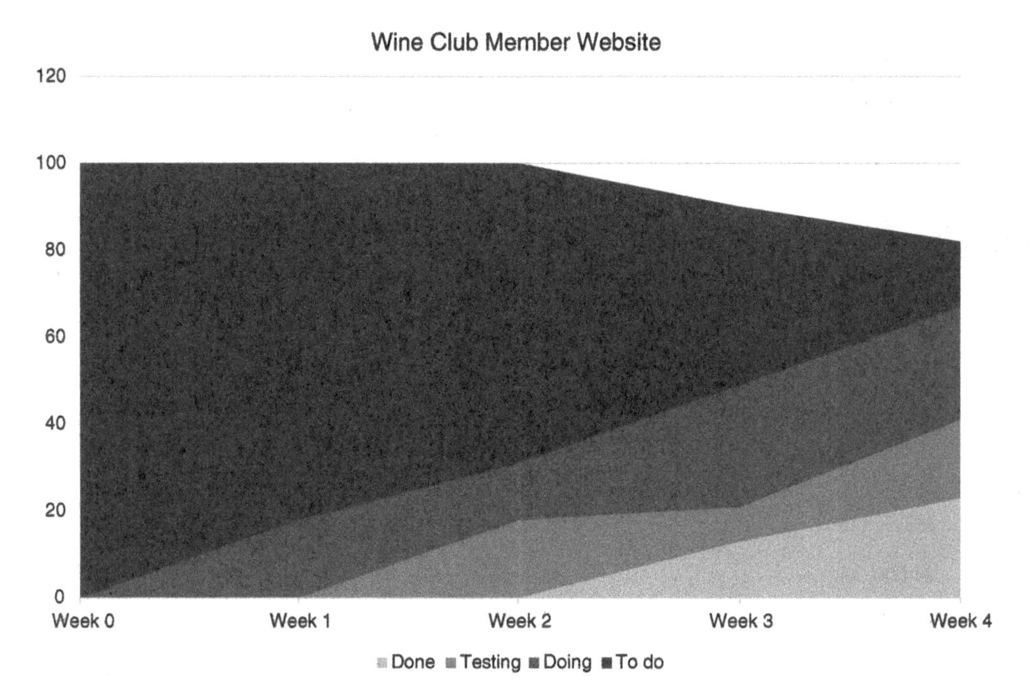

FIGURE 21-5 Stacked area chart.

Bar Charts

Bar charts shows the relationship between one or more sets of data. They are good for analyzing the performance of categories of data. You can show vertical or horizontal bars. You can create bar charts that are clustered, stacked, or 100% stacked. Figures 21-6, 21-7, and 21-8 show different bar charts, all based on the same data. You can see the different presentations based on a survey the Dionysus Winery conducted to find out what types of activities potential guests would enjoy. These charts show the results of surveying 50 potential guests in three different age groups.

Even though all the charts used the same data, the presentation for the stacked bar chart more clearly shows the preference for wine making activities.

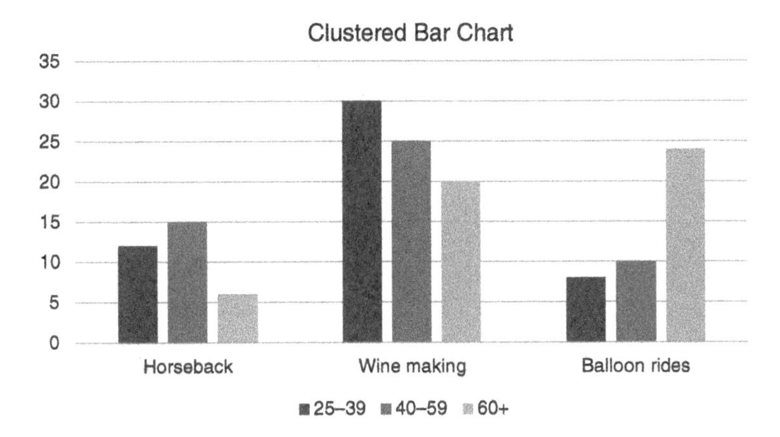

FIGURE 21-6 Clustered bar chart.

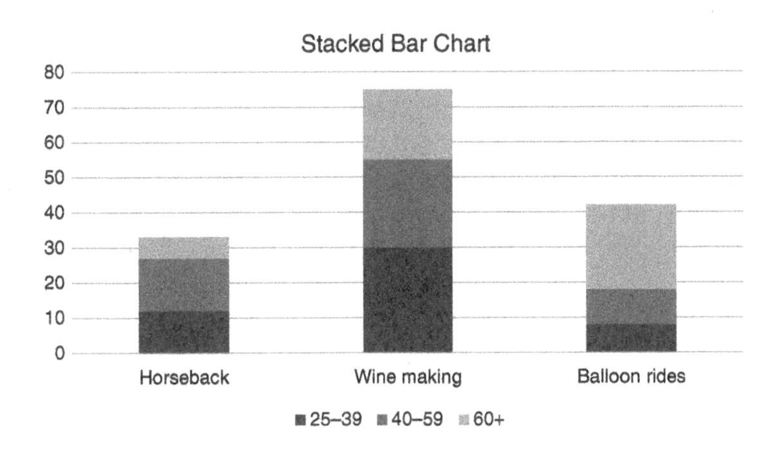

FIGURE 21-7 Stacked bar chart.

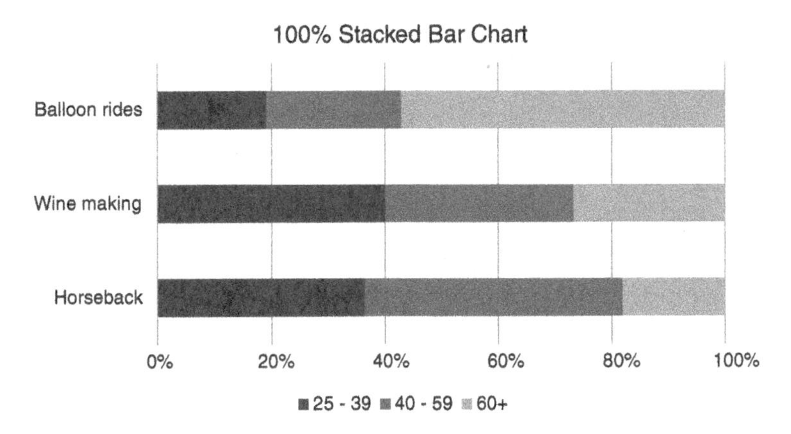

FIGURE 21-8 100% stacked bar chart.

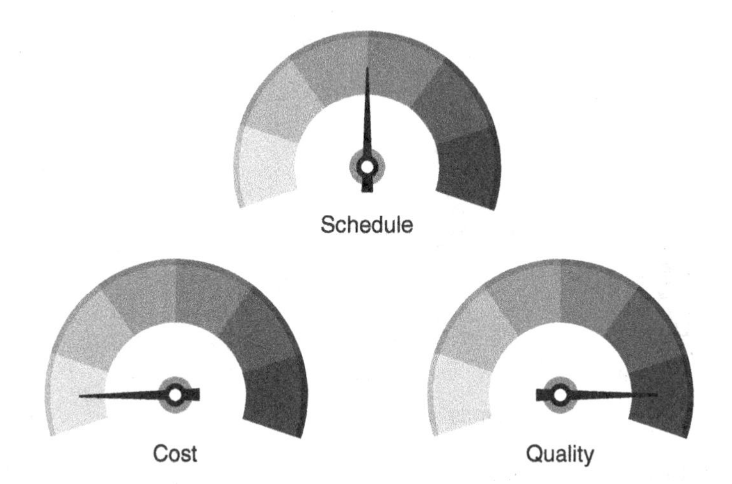

FIGURE 21-9 Gauges.

Gauges

Gauges assess values on a relative scale; they are like your gas gauge or speedometer. They show a point in time measure rather than a trend. You can use them to indicate the degree of risk, cost variance, customer satisfaction, team mood, and other relative values. Figure 21-9 shows a set of gauges.

Scatter Diagrams

Scatter diagrams show relationships between two variables, such as cost and customer satisfaction, or risk probability and impact. Figure 21-10 shows the relationship between the room rate and expected customer satisfaction based on a survey of potential customers for the Dionysus Winery.

This demonstrates a negative correlation, meaning that as one variable increases, the other decreases. In this case, as price increases, satisfaction decreases. You can see it

FIGURE 21-10 Scatter diagram.

isn't a precise linear correlation: There is a big drop in satisfaction when the price goes from $250 per night to $300.

Bubble Charts

Bubble charts show three variables. Their placement on an X-axis, Y-axis, and the size or color of a bubble can be used to relate performance information. The bubble chart shown in Figure 21-11 shows earned value numbers. The X-axis shows the schedule performance index (SPI), the Y-axis shows the cost performance index (CPI), and the size of the bubble indicates the cost-schedule performance index (CSI). The CSI is calculated by multiplying the CPI times the SPI.

Remember, with earned value, the farther your indexes are from 1.0, the more concerning the performance. Here can see there are two work packages that have a CSI of .77. These are the two that you need to pay the most attention to. Conversely, those in the upper right corner are of the least concern.

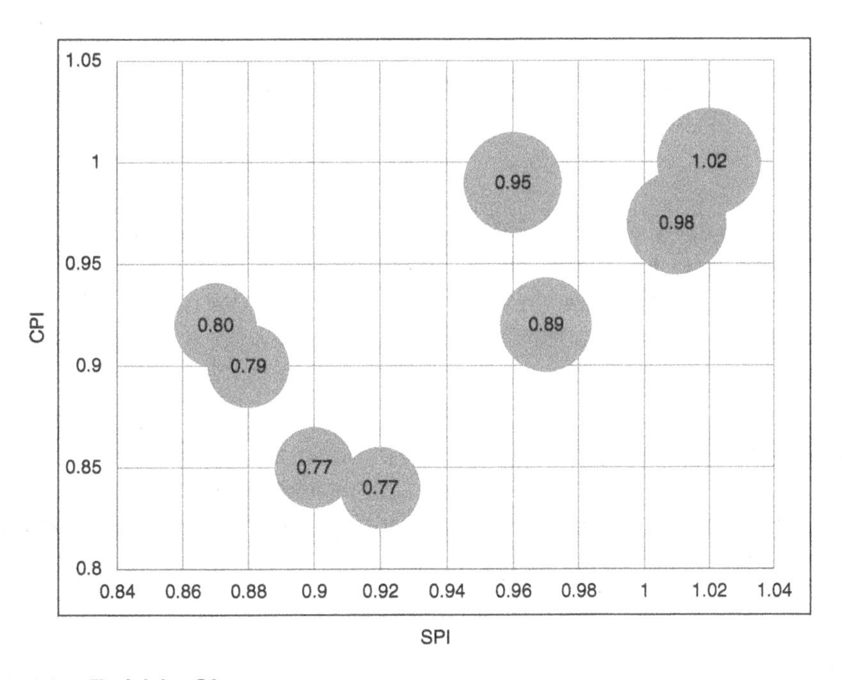

FIGURE 21-11 Bubble Chart.

Pie Charts

We are probably all familiar with pie charts. Pie charts show relative percentages of multiple items. They provide a great visual image of the relative value of the variables. However, they are only useful if there are less than six elements; otherwise,

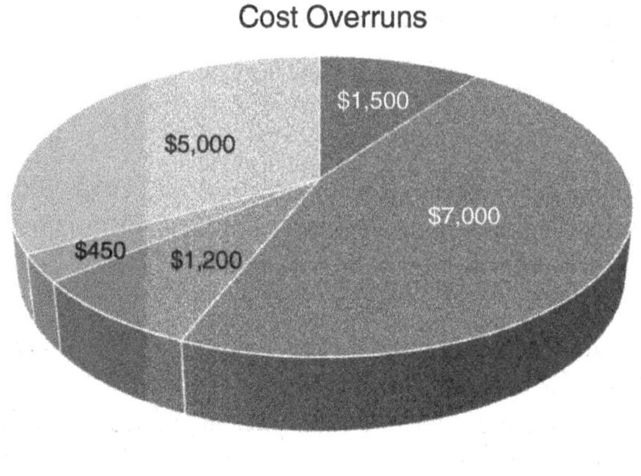

FIGURE 21-12 Pie Chart.

they get too crowded, and it is difficult to differentiate small value differences. Figure 21-12 shows a pie chart that identifies the amount of cost variance for construction work packages.

Trend Icons

Trend icons indicate if a measure is moving toward the target, staying the same, or moving away from the target. You can integrate these in with other measures, such as gauges or line charts. Trend icons can include a thumbs up 👍 or down 👎.

They might show arrows that point up⬆, down⬇, sideways➡, or diagonally⬈.

Radar Diagrams

Radar diagrams are also known as spider charts or star charts. They are used to compare multiple options or variables against multiple characteristics. In Chapter 2 we used a radar diagram to assess the best development approach to use for three different projects by comparing 15 different variables. Figure 21-13 shows the same diagram shown in Chapter 2.

With so many options it can be hard to know which type of chart is best to show the information you want to communicate. To help with that, Table 21-4 shows a summary of different types of charts and the purpose for which they are best used.

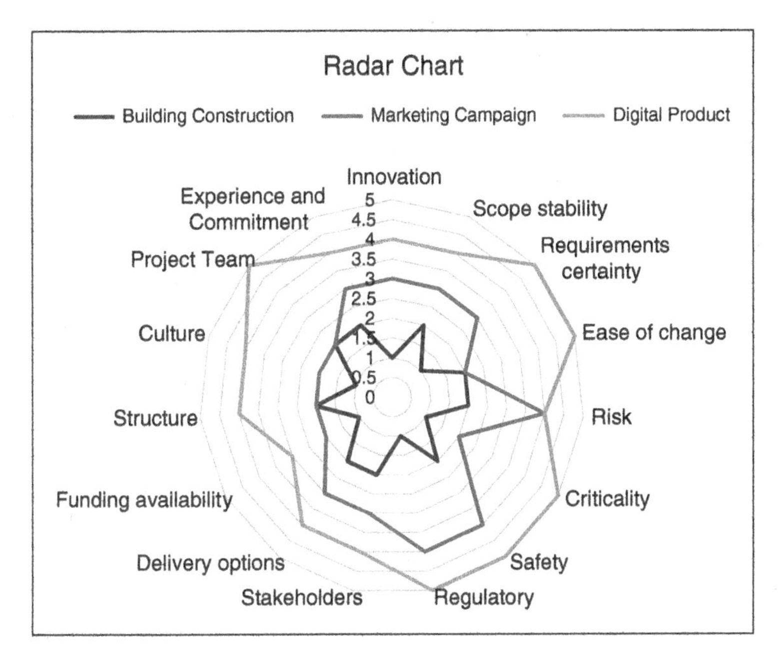

FIGURE 21-13 Radar chart.

TABLE 21-4 Chart Summary

Traffic light	Traffic light charts provide a top-level view of the status and indicate if performance is approaching a threshold.
Line charts	Line charts show changes and trends over time.
Area charts	Area charts are similar to a line chart, but they use color or shading to communicate information.
Bar charts	Bar charts show the relationship between one or more sets of data.
Gauges	Gauges assess values on a relative scale.
Scatter diagrams/ bubble charts	Scatter diagrams show relationships between two variables. Bubble charts add a third variable with the size or color of the bubble.
Pie charts	Pie charts show the relative percentages of multiple items; use them for six or fewer items.
Trend icons	Trend icons indicate whether a measure is moving towards the target, staying the same, or moving away from the target.
Radar diagrams	Radar diagrams compare multiple items against multiple characteristics.

INFORMATION RADIATORS

Information radiator: A collection of information displayed openly so team members and other stakeholders can see the status of the project.

Many Agile projects use information radiators to convey status information. An information radiator is an electronic display or hand drawn chart that is placed in a visible location so anyone can easily see the current status of the project. When the information is handwritten or drawn it may be referred to as a "big visible chart."

We have discussed different information radiators in previous chapters. The following list summarizes that information.

- **Burndown chart** (Figure 20-1): Estimates remaining work by showing a target completion rate of all the work. The team tracks actual completed work against the target completion rate. This indicates if work is ahead or behind the target completion rate.
- **Burnup chart** (Figure 20-2): Shows the work accomplished and compares it to the target completion rate.
- **Task board** (Figure 20-7): Shows the status of tasks from backlog to completion. Also identifies bottlenecks in the process if there are too many tasks in one column.
- **Mood charts** (Figure 20-10): Use emojis to indicate the mood or attitude of team members.
- **Retrospective charts** (Figures 11-1 and 11-2): Used to improve the team process and outputs. They categorize outputs from a retrospective meeting and give visibility into team feedback on what is working, new ideas to try, and so forth.

Figure 21-14 shows an example of how you can display information from several charts.

HYBRID DASHBOARDS

For a hybrid project you will need all the predictive performance information and all the Agile information in an electronic format. To determine the contents, you should meet with stakeholders to identify the information that is most important to them. You can also find out if they want the report pushed out to them, or if they want to be able to log onto a portal to see the latest information whenever they want. Some dashboards have the ability to send out alerts if a specified threshold has been crossed, such as when a variance moves from yellow to red status on a stoplight chart.

You should also establish expectations of how frequently the information in the dashboard will be updated. You will need to weigh the benefit of frequent updates with the work it takes to keep everything updated. It might be easy to update some information daily, such as a task board or burn chart. Other information is dependent on enterprise

Team Member	Monday	Tuesday	Wednesday
Issa			
Emily			

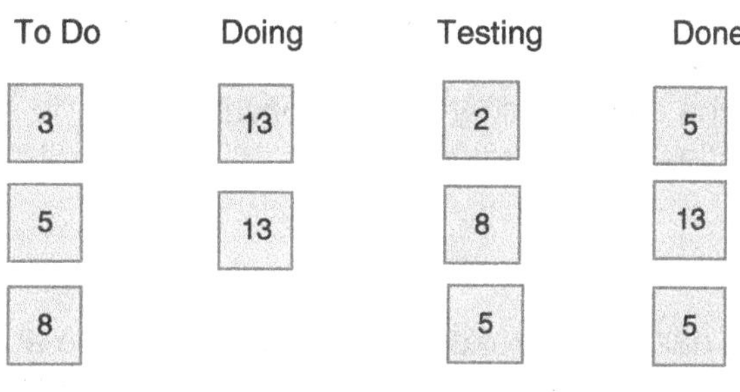

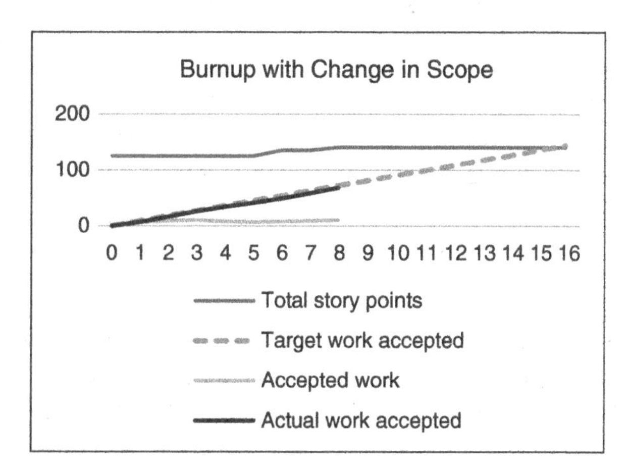

FIGURE 21-14 Sample Agile Information Radiators.

information, such as budget disbursements or work performed by contractors. Earned value information is typically updated monthly, though for large projects it may be updated more frequently due to the significant costs involved.

Robust dashboard systems allow you to drill down into the details behind the metrics so you can find out which tasks are causing the schedule to be late, or whether it is labor or materials that are driving a cost variance.

Tips

There are a few tips to keep in mind for creating useful dashboards. The first is to keep it simple and keep it clean. The beauty of a dashboard is it imparts a lot of information in an easy-to-absorb format. However, if it is too cluttered, it is not easy to absorb. Only select the most relevant information for your dashboard. That doesn't mean you can't get data and information on performance that is not on a dashboard, it just means don't put everything on a dashboard.

Another tip to consider is the presentation. While the data itself is paramount, the way it is presented makes a difference. Make sure your colors are telling the story you want (don't use red to indicate good performance) and that they are complementary. You don't want your dashboard to be jarring to look at. Consider the size of the charts—make sure they are readable, that the different charts are relatively the same size, and that you keep the legend placement, titles, and fonts consistent.

Benefits

Establishing a reporting dashboard takes an investment of time and energy, but the benefits are many. Chief among them is that they visually display a large amount of information in an easy-to-absorb presentation. The charts and graphs provide an overview of the project status at a glance. This allows you to spot potential problems and negative trends easily so you can make timely, informed decisions.

With all these benefits, they do have a few limitations. A dashboard does not tell you what caused variances or negative trends, nor do they provide contextual information. You need to look to a narrative report for that type of information.

SUMMARY

In this chapter we described narrative reports and visual reports. Narrative reports include status reports, variance reports, and earned value analysis reports. Visual reports include dashboards and information radiators. Dashboards provide an at-a-glance view of key performance indicators. They use multiple charts to show performance for scope, quality, schedule, cost, and other indicators. Information radiators are used for Agile deliverables. They ensure team members and other stakeholders can see the status of the project. Hybrid dashboards can include both predictive and adaptive information.

Key Terms

information radiator

22

Corrective Actions and Closure

Throughout your project you will find yourself needing to respond to potential and real problems. Some of the problems will be surfaced through reports, some will be the result of risk events occurring, some will be logged as issues in an issue log, and still others will seem to appear out of thin air. To work with these effectively you will need to implement preventive actions to keep performance on track and corrective actions to realign performance with expectations.

In this chapter we will identify a process for applying preventive and corrective actions. We will describe some of the causes of variances along with potential responses. We will also identify when it is appropriate to re-baseline. Because all good things must come to an end, we will discuss transitioning the project at closure, along with importance of acknowledgment and close-out reports.

PREVENTIVE AND CORRECTIVE ACTIONS

Based on the information you discover in your reports, you will need to determine the best action to take. You can follow this four-step process to identify and implement corrective actions.

1. **Analyze the problem:** This includes conducting a root cause analysis to determine the source of the variance or other performance issue. This can include identifying the source of quality defects, schedule delays, or budget overruns.
2. **Compare potential responses:** You may identify multiple options for correcting a variance. When looking at potential solutions, keep in mind that not all variances need to be addressed. Some variances are within the threshold of acceptable performance, and other variances you won't be able to do anything about. Before taking action, ask yourself, If the variance continues, will it be a problem? If not, note the variance, and move forward.
3. **Select the best response:** Once you have determined the best response, you may need to gain authorization and commitment to implement it. Some responses entail spending reserve funds, bringing in additional resources, or getting outside help. These usually require buy-in from the sponsor or other stakeholders. Make sure you have buy-in and agreement before you take action.
4. **Implement and review:** Implement the response, and then check to see whether the response effectively addressed the issue. You may need to take additional actions if your response did not correct the problem.

Potential Causes and Responses for Performance Issues

Often there isn't just one cause for a performance variance; for example, a schedule delay for one task can create a problem for future tasks if resources that were scheduled for the future work aren't available based on the schedule delay. This can cause further delays while trying to find available resources with the right skill sets. Finding resources on short notice can result in higher rates. Thus, you can see there can be multiple sources of a performance issue.

In the next sections we will look at possible causes and potential responses for scope, schedule, and budget problems.

Scope and Quality

Scope issues can be the result of poorly defined requirements, changing requirements, a realized risk, or other causes. To address problems with scope and quality, consider the following options:

1. Clearly define acceptance and completion criteria. If the acceptance and completion criteria for a deliverable is not clear and agreed upon, there can be a difference of opinion on what is considered "done."
2. Employ validation and verification activities as described in Chapter 6. Verifying performance and getting validation from the customer throughout the process reduces the likelihood of delivering an end product that doesn't meet expectations.

3. Track any defects and identify the root cause. Keep a defect log so you can identify defects early in the process and fix them rapidly.
4. Reprioritize the backlog. For Agile development, you may be able to reprioritize scope so you can continue to deliver value. This may buy time to resolve the issue. You can also reprioritize requirements using a method such as MoSCoW, where:
   M = Must have;
   S = Should have;
   C = Could have;
   W = Won't have.
5. Consider de-scoping. If you have a component or deliverable that is causing problems and it is not critical to the project, consider eliminating it.

Schedule

Schedule issues can be the result of scope issues. They can also occur because of unrealistic expectations or unrealistic estimates. Lack of skilled resources is also a source of schedule delays. To address scheduling issues, consider the following options:

1. Crash activities on the critical path. In Chapter 8 we described crashing as compressing the schedule by adding resources—such as people, or money. This entails making cost and schedule trade-offs, such as bringing in additional resources, paying overtime, or expediting work.
2. Fast-tracking activities is compressing the schedule by performing work in parallel that is usually performed sequentially. This includes overlapping some activities, inserting a lead between activities, or changing dependency types. You can review information on fast-tracking in Chapter 8.
3. Reprioritize scope, or descope as described above.

Budget

Budget issues can be the result of scope issues or missed requirements. Schedule delays or crashing the schedule can also lead to cost variances. Material variances, either in amount or price, can lead to budget overruns. Team members using overtime, more hours than estimated, or having higher rates can also cause budget variances. Unrealistic expectations or poor estimates are another cause of budget issues. To address budget variances, consider the following options:

1. Reassess materials: See whether there are ways to use less materials and supplies or less costly materials and supplies.
2. Reassess team members: Determine whether you can use fewer people or people with lower hourly rates. If you are paying for a highly skilled resource but don't need that level of skill, you can probably get the job done for less.

3. Analyze expenditures: Look to see whether you are being charged for overtime or whether there are other expenses you did not authorize.
4. Reprioritize scope or descope as described in the section titled "Scope and Quality" above.

Updating the Baseline

Sometimes you can't correct a variance. In those cases, you will need to re-baseline. Re-baselining involves changing one or more baselines (scope, schedule, or cost). You should not re-baseline to mask a variance or if the variance is minimal to moderate. In those cases, it is appropriate to show the variance as part of the project performance. However, there are a few situations where re-baselining is appropriate:

1. A change in scope or requirements;
2. A significant change in resources;
3. If the schedule is shortened or lengthened;
4. If the budget is reduced or increased;
5. When the schedule or cost variance is so great that continuing to measure against it doesn't really provide meaningful information.

When you re-baseline, it is important to keep all the information from the previous baseline. You will need this for reference and for documentation. You should note the date that you switch from the old baseline to the new baseline so you have context for why some measures suddenly went from red to green. It is also a good idea to document what changed from the previous baseline for historical reference.

PROJECT CLOSURE

All good things must come to an end, even hybrid projects. With that in mind, it is important to plan for closure from the start of the project. Project closure should address the following:

- Transitioning the final product or service;
- Performing administrative closure work;
- Acknowledging your team;
- Evaluating success;
- Submitting close out reports.

We'll look at each of those next.

Transition

Even when you are planning your project, you should be thinking about transitioning to the client or operations. The people who will be inheriting and operating the end product or result can provide valuable input on the best way to build or package deliverables so they are easy to maintain and operate.

When you transition the final deliverables, make sure you have provided the people receiving the end product with the necessary training and documentation. You may need to do some on-the-job training or have your technical team members available for questions for the first 30 days of operation.

Administrative Closure

Administrative closure is about making sure the physical, administrative, and contractual aspects of the project are complete. For the physical space, make sure that all furniture, equipment, supplies, and materials are clean, catalogued, and distributed appropriately.

Administrative issues include making sure that all project documentation is in order. Some industries have regulations that require organizations to keep documentation for a specified number of years. You will want to make sure all project records are organized and archived, preferably electronically, so they can be retrieved easily if needed. Future projects may want to learn from your schedule, budget, risk register, assumption log, and so forth.

For projects that outsourced part of the work or that had vendors or contractors, there are some tasks you will need to complete. In most organizations, the project manager does not have contractual authority. A legal or purchasing department has that authority. The project manager and the contracting representative will work together throughout the project to identify contractors, develop a statement of work, and eventually close out the contract. As a project manager you will want to inform the contracting representative that the work is complete, let them know whether you think the contractor or vendor should be used in the future, and advise them that they can close out the contract. You should also work with accounts payable to make sure that all invoices are scheduled for payment.

Acknowledgment

As mentioned earlier in this book, your most important resource is your team. The end of a project may be a relief, or it may be a bittersweet experience if you had a really great team. It is always nice to acknowledge the team as a whole for accomplishing the work. If you have the budget for it, it is great to have an appreciation event, such as a nice meal or a fun event.

If your team is not too big, it is nice to recognize individual contributions as well. You can write letters for their personnel file, write personal notes, buy little gifts, or find some other way to acknowledge and appreciate their contribution. Saying "thank you" is a bare minimum, but it is important. Make sure your team feels appreciated for the work they did.

Evaluating Success

In Chapter 1 we noted that the purpose of projects is to bring value to stakeholders. At the end of the project, you should evaluate whether you accomplished that goal. Sometimes you won't know until well after the project is complete, but in most cases, you can evaluate whether the project was successful.

The first place to look to determine whether the project was successful is the project charter. The charter identifies high-level requirements, project objectives, and success criteria. You should revisit those to see whether you fulfilled all the requirements and met the success criteria. Obviously, part of success is timely delivery. Did you meet all your milestones? Was the final delivery on time? Another important aspect is delivering on budget or within an acceptable variance.

One aspect of project success that is sometimes overlooked is stakeholder and team member satisfaction. If you met all the acceptance criteria but your stakeholders are not satisfied, is that considered a success? Likewise, if you delivered on time and on budget but your team is burned out and doesn't want to work with you again, you probably can't consider the project a complete success.

The results of your evaluation will be recorded in your final project report and perhaps in your final lessons learned report.

Close-Out Reports

There are two close-out reports, a final project report and a final lessons learned report. The final project report is a summary of the project performance. The final lessons learned report compiles information from the lessons learned and retrospectives that were undertaken throughout the project.

Final Project Report

This report should summarize high-level information about the project. It can be very detailed or quite brief, depending on the project. The purpose of this report is to provide the sponsor with a summary of the project and to record information for historical purposes. Table 22-1 provides an outline for a final project report.

This is just an outline. Depending on your project, you will want to tailor this content to meet the needs of your project and your organization.

TABLE 22-1 Final Project Report

Project description: Provide a summary level description of the project; this information can be copied from the project charter.

Performance Summary

Scope: Describe the scope objectives and the success criteria for the project; provide evidence that the success criteria were achieved; describe any variances, along with the reason and impact for the variances. If appropriate, differentiate between adaptive work and predictive work.

Quality: Describe the quality requirements and acceptance criteria for the deliverables; summarize the verification and validation information for deliverables; describe any waived requirements or variances, along with the reason and impact.

Schedule: Document the specific dates that needed to be met to meet the schedule objectives, including milestone delivery dates. Document the actual dates and describe any variances along with the reason and the impact for the variances. If appropriate, differentiate between adaptive and predictive scheduling.

Cost: Enter the approved budget for the project, along with the range that indicates budgetary success; enter the final project costs; describe any variances, along with the reason and impact for the variances.

Risks, Issues, and Impediments

Provide a summary of the risk register, issue log and impediment log; discuss any significant risks, issues, or impediments that affected the project, along with the responses.

Final Lessons Learned Report

A final report for lessons learned on a hybrid project will summarize information collected from lessons learned throughout the project as well as relevant information from team retrospectives. Often the team comes together at the end of a project to conduct a lessons learned meeting, and the information from that meeting is captured in the final lessons learned report. The purpose is to identify those aspects of the project that went well, so other teams can adopt those practices. You also want to cover aspects of the project that didn't go well and discuss how that can be avoided in the future.

Here is a list of topics you can include in your report. For each topic, think about practices that were effective and practices that can be improved. You will want to tailor the list to meet the needs of your project.

- **Team performance:** Consider both adaptive team dynamics and the team as a whole.
- **Scope evolution and management:** Describe how scope evolved for the adaptive deliverables and how well it was controlled for the predictive deliverables.
- **Schedule performance:** In addition to overall schedule performance, describe how well the adaptive parts of the project integrated with the predictive parts of the project.

- **Cost performance:** Consider how well the budget was managed with evolving scope and predictive scope.
- **Estimating:** Note techniques that worked well for estimating duration, velocity, resources, and costs.
- **Quality processes:** Discuss processes used for adaptive and predictive deliverables and how effective they were. If appropriate, you can include more detailed information on quality defects.
- **Stakeholder engagement:** Note how effective stakeholder engagement was using adaptive practices, and how effective it was for the predictive deliverables.
- **Risk, issue, and impediment management:** Discuss the process for risk, issue, and impediment management. If appropriate, you can include more detailed information on significant events or challenges.
- **Other:** Describe any additional information on areas that worked well and areas that could be improved.

SUMMARY

Depending on the information gleaned from the reports, you may need to take preventive or corrective action for scope/quality, schedule performance, and cost variances. We reviewed various causes for these performance issues and potential responses. In some cases, such as if there is a scope change or the variance is severe, you may need to re-baseline.

We described how closing the project entails transitioning the final product or result to operations, along with performing administrative and contractual closure activities. Acknowledging your team is an important and a fun part of closing out the project. Finally, we described two important close-out reports, a final project report and a final lessons learned report.

Key Terms

There are no new terms in this chapter.

23

Making the Move to a Hybrid Environment

Making the move to a hybrid environment is not a one-size-fits-all endeavor. Each organization is different, but there are some things that all organizations can consider when moving toward hybrid project management. In this chapter we will provide a brief overview of some of the considerations for helping organizations successfully transition to a hybrid way of practicing project management.

ESTABLISH CRITERIA

Hybrid project management is all about choosing the best approach for delivering value. With hybrid project management you have a range of options to select from, but you should have some criteria or guidelines to help determine the best approach. In Chapter 2 we provided 15 variables to consider when determining the best approach for development methods. These are a good starting place to identify criteria that point to purely predictive approaches, purely adaptive approaches, and mixing and matching with a hybrid approach.

When implementing these criteria, keep in mind that things will shift and change. The criteria you start out with might change as people get more comfortable with different approaches or as you find that some criteria need to be tweaked to get better outcomes.

One thing that is important to keep in mind, there are few hard-and-fast rules when it comes to development approach. Just because part of your project includes IT, that doesn't mean you have to use an Agile approach. And for those aspects of your project that are not IT, it doesn't mean you can't use an Agile approach.

ESTABLISH THE RIGHT ENVIRONMENT

Making a commitment to a hybrid environment takes preparation. The most critical success factor is making sure you have support of your senior leaders. Once they are onboard, you need to make sure team members, project leads, scrum masters, sponsors, and product owners are all aligned with the move to a hybrid environment.

Once everyone is aligned, you will want to establish the appropriate governance. This means balancing structure with the ability to be flexible and responsive to change. You may establish a project management office (PMO), value delivery office (VDO), or some combination of the two. Make sure it fits your needs.

With governance comes policies, terminology, practices, and templates. You'll need to determine whether you are doing "sprint planning," "iteration planning," "scope evolution," and so on. Everyone needs to use the same language and be clear on what it means. There should be alignment with practices such as daily stand-ups and weekly status meetings. All of these can evolve as you discover what works and what doesn't, but you should start out with some guidelines and then evolve them as needed.

The metrics and measures you use to monitor progress for predictive projects are different than those you use for adaptive projects. Make sure that both sets of metrics remain focused on the end goal, which is delivering value and meeting project objectives.

As you get ready to shift, remember that it is fine for some projects to maintain purely adaptive practices and some to maintain purely predictive practices. Hybrid is about making things work!

PROCESS FIRST

When some organizations are getting ready to adapt a new set of practices or processes, the first thing they do is find a system or software that can support those practices. This is great, but acquiring a new software application should be the last step, not the first! Before you invest in software, spend time figuring out what you need it to do. This means agreeing to the following first:

- Terminology;
- Roles;
- Processes.

Once those things are in place, you can think about training. You will likely have some team members who aren't familiar with Agile practices. You may have some that are only familiar with Agile and aren't skilled in reading a predictive schedule.

Once all these things are in place, you should have a better idea of your requirements for a software tool. Software vendors can provide a demonstration so you can evaluate how closely their software aligns with your needs. They can also provide training on the tool and the templates that are included with the software.

Hybrid project management gives project leaders the freedom to choose how to deliver value. I hope that whether you are new to project management or you have decades of experience, you have found some new ways of practicing and discovered some new techniques you can apply on your projects.

Glossary

actual cost (AC): The funds spent to accomplish work.

adaptive: An approach for creating deliverables that allows for uncertain or changing requirements.

affinity grouping: Categorizing elements into groups based on similar characteristics.

Agile: An adaptive way of delivering value by following the four values and 12 principles established in the Agile Manifesto.

analogous estimating: Using information from previous similar projects to develop an estimate for the current project.

assumption: Something that is considered true for purposes of the project.

assumption log: A dynamic document that is used to identify and track assumptions and constraints.

authority: The right to make and approve decisions.

backlog: A list of work to be done.

baseline: An agreed upon version of a document that is used to measure progress and detect variance.

bias: An opinion or feeling that is preconceived and unreasoned.

bottom-up estimating: Summing estimates from individual work packages to arrive at an overall estimate.

budget: The approved time-phased estimate for project work.

budget at completion (BAC): The total cost for the work to be done.

buffer: Time inserted into the schedule to protect delivery dates.

burndown chart: A graph that maps the actual work completed to the target work to be completed.

burnup chart: A graph that shows the total work along the top and a line that shows the work completed.

change management: A process for identifying, tracking, evaluating, and implementing modifications for deliverables, plans or project documents.

change control board: A group that reviews, analyzes, approves, or rejects changes to the product or the project.

competency: The skill level of an individual.

configuration management: A system used to identify and track individual items in a product or release.

constraint: A limiting or restricting factor.

contingency reserve: Funds or time incorporated into a baseline to accommodate known risks and risk responses.

control account: A component of the WBS (work breakdown structure) used to control work performance and report on cost and schedule status.

convergence: A point in a schedule where multiple paths merge.

cost performance index (CPI): A measure of cost efficiency calculated by the earned value divided by the actual cost (EV/AC).

cost variance (CV): The difference between the actual costs and baseline costs. In earned value management, the difference between the earned value and the actual cost (EV – AC).

crashing: Compressing the schedule by adding resources, such as people or money.

critical path: The series of tasks through a project with the longest duration, which determines the soonest the project can be completed.

critical path methodology: A scheduling approach that identifies the path that drives project duration and identifies the amount of scheduling flexibility on other paths.

criticality: The importance of a component, deliverable, or project.

cumulative flow diagram: A chart that shows the flow of work from backlog, through different stages of development, to completion.

cycle time: The time it takes to get a piece of work through a part of the development process.

daily stand-up: A brief meeting held by adaptive teams to review progress from the previous day, describe the intention for the day's actions, and surface impediments.

decision matrix: A tool used to evaluate multiple options against a set of criteria.

decision tree analysis: A diagramming technique for evaluating multiple options in the presence of uncertainty.

deliverable: A component or subcomponent of a product or service. A deliverable can be stand alone, or part of a larger deliverable.

development approach: The means by which the project team will create and evolve deliverables.

divergence: A point in a schedule where one task separates into multiple paths.

duration: The amount of time needed to complete work.

earned value (EV): The value of work accomplished.

earned value management (EVM): A method used to plan and track project performance.

effort: The amount of labor needed to complete work.

elicitation: A structured approach to draw out requirements.

emotional intelligence: Awareness of one's own emotions and moods and those of others, especially when leading people.

estimate at completion (EAC): The expected total cost of the project.

estimate to complete (ETC): The expected cost to finish the remaining work.

expected monetary value: A statistical technique that calculates the average outcome when the future includes multiple potential outcomes.

fallback plan: A risk response used when other responses have failed.

fast-tracking: Compressing the schedule by performing work in parallel that is usually performed sequentially.

feeder buffer: Time inserted into the schedule where a path merges with the critical path.

Fibonacci sequence: A series of numbers in which a number is the sum of the two preceding numbers.

finish-to-finish: A relationship where the preceding task finishes before the next task can finish.

finish-to-start: A relationship where the preceding task finishes before the next task can start.

float: The amount of time a task can slip and not affect a project constraint or the end date. Also known as total float.

free float: The amount of time a task can slip and not affect the following task.

Gantt chart: A bar chart used in scheduling where tasks are listed in rows and bars indicate the duration of the tasks.

generalizing specialists: People who have deep knowledge in one area and broad knowledge or skills in complementary areas. *Also known as* T-shaped people.

hybrid project management: A blend of predictive and adaptive approaches to delivering value, determined by product, project and organizational variables.

I-shaped people: People who have deep knowledge in one specific area.

impediment: Anything that gets in the way of the team achieving its objectives.

incremental: An adaptive development approach that begins with a simple deliverable and then progressively adds features and functions. *See also* adaptive.

information radiator: A collection of information displayed openly so team members and other stakeholders can see the status of the project.

integrated master schedule: A schedule that combines all schedules for a project into one all-encompassing document.

issue: A current condition that may have an impact on a project.

iteration: A brief, set time interval in a project where the team performs work. *Also known as* timebox or sprint.

iteration plan: A documented approach the team will use to accomplish the work in a timebox. *Also known as* sprint plan.

iteration planning: An event to identify, clarify, and estimate work that will be done in the current iteration.

iteration review: An event to demonstrate the work that was accomplished in the current iteration.

iterative: An adaptive development approach that begins with delivering something simple and then adapts based on input and feedback. *See also* adaptive.

known risks: Risks you can anticipate and plan for.

lag: A delay between tasks.

lead: An acceleration between tasks.

lead time: The time it takes to get a piece of work through the entire development process.

lessons learned meeting: A meeting to identify and document areas that are working well on the project and areas that can be improved.

management reserve: Funds or time that is not part of a baseline. Used to address unplanned, in-scope work.

minimum viable product: The first release of a product that contains the least number of features or functions in order to be useful.

mood chart: A visual representation of team members' moods or attitudes.

multipoint estimating: Developing optimistic, pessimistic, and most likely estimates and calculating an average or weighted average.

Net Promoter Score®: A measure of stakeholder loyalty and satisfaction.

network diagram: A visual depiction of the schedule using nodes and arrows.

opportunity: A risk that will have a positive impact on a project.

parametric estimating: Developing a mathematical model based on significant historical data.

performance measurement baseline (PMB): A baseline developed by integrating scope, schedule, and cost information.

phase gate: A review at the end of a phase to determine whether the project is ready to advance to the next phase.

planned value (PV): The value of work expected to be accomplished.

planning package: A component in the WBS that represents work that hasn't been decomposed into work packages.

point estimate: A single value that represents the best prediction for an outcome.

predictive: An approach for creating deliverables that seeks to define the scope, schedule, and budget toward the beginning of the project and minimize change throughout the project. *See also* waterfall.

product owner: The person accountable for the performance of a product.

progressive elaboration: Developing more detail as more information is known about the project.

project buffer: Time inserted into the schedule before the final delivery date.

project charter: A document that formally authorizes a project and provides a high-level description of the project.

project life cycle: A series of phases a project goes through from inception to completion.

project management plan: A plan that describes how the project will be planned and managed.

project manager: The person accountable for leading the team to deliver the project outcomes.

project roadmap: A graphic high-level view of the project that includes phases, reviews, milestones, and other key information.

project vision statement: A brief and compelling statement that describes the future state after the project is complete.

RACI chart: A type of responsibility assignment matrix that identifies who is responsible, accountable, consulted, and informed.

radar chart: A graphical chart that displays multiple quantitative variables on numeric axes. *Also known as* spider chart, web chart, or star chart.

release: A set of features, functions, or deliverables that are placed into use.

release plan: A plan that shows the expected timing, milestones, and outcomes for releases.

requirement: A capability that must be present or a condition that must be met to achieve the project objectives.

requirements traceability matrix: A grid that links requirements to deliverables and other project elements.

reserve: Additional funds or time added to a baseline to account for cost or schedule risk.

resilience: The ability to adjust or recover readily from adversity, crisis, setbacks, change, and other significant sources of stress.

resource breakdown structure: A hierarchy chart that shows resources grouped by type or category.

resource histogram: A bar chart that shows information about resources, such as number or skills.

resource leveling: Reallocating resources so no instances of overallocation remain.

resource loading: Entering resources into a scheduling tool.

resource smoothing: Reallocating resources within the available float, so the critical path is not affected.

responsibility: The work a person in a role is expected to perform.

responsibility assignment matrix (RAM): A chart that shows the role team members have on a work package.

retrospective: A workshop to review the work product and processes to find ways to improve outcomes.

risk: An uncertain event or condition that can have an impact on a project.

risk acceptance: Acknowledging the existence of a risk, but not taking action unless the risk occurs.

risk-adjusted backlog: A backlog that incorporates actions to reduce impediments or threats to the project.

risk avoidance: Taking action to eliminate the threat.

risk breakdown structure (RBS): A hierarchy of potential sources of risk.

risk escalation: Bringing the risk to someone with more authority to address it.

risk management plan: A subsidiary plan of the project management plan that describes how risks will be identified, analyzed, and addressed.

risk mitigation: Reducing the probability and/or impact of a threat.

risk register: A log for documenting information on threats and opportunities.

risk threshold: The amount of risk exposure an entity is willing to accept before actively addressing a risk.

risk tolerance: The degree of uncertainty an entity is willing to accept.

risk transfer: Shifting the management and response of a threat to a third party.

risk trigger: An event or condition that indicates a risk has occurred or is about to occur.

role: A position on a team that describes the type of work the person does.

rolling wave planning: A form of progressive elaboration where activities in the near future are planned in detail, and activities farther out in the future are kept at a high level.

schedule compression: Reducing the schedule duration without reducing scope.

schedule management plan: A subsidiary plan of the project management plan that describes how the schedule will be developed, managed, and maintained.

schedule performance index (SPI): A measure of schedule performance efficiency calculated by the earned value divided by the planned value (EV/PV).

schedule variance (SV): The difference between the planned accomplishments and actual accomplishments. In earned value management, the difference between the earned value and the planned value (EV – PV).

scope creep: An increase of product or project scope without the necessary cost or schedule increase.

scope management plan: A subsidiary plan of the project management plan that describes how scope will be defined, documented, managed, and verified.

scope statement: A document that describes project scope and deliverables and identifies out of scope work.

scrum master: The person who supports the team in maintaining alignment with Agile values and principles.

secondary risk: A risk that occurs as a result of a risk response.

servant leadership: A leadership style that focuses on the needs of the team.

self-managing teams: Teams that share accountability for the delivery of value.

sponsor: The person who provides project resources and supports the project manager in meeting the project objectives.

stakeholder: People and groups who can affect your project or are affected by your project.

stakeholder register: A document that records relevant information about project stakeholders.

start-to-start: A relationship where the preceding task starts before the next task can start.

status meeting: A meeting to discuss the current progress on the project.

story points: A relative unit of measure used to estimate work in a user story.

subsidiary plan: A component of the project management plan that describes how a specific aspect of a project will be planned and managed.

summary task: A task that aggregates information from detailed work into a single task.

system: A set of interacting, interrelated or interdependent elements, forming a whole.

T-shaped people: People who have deep knowledge in one area and broad knowledge or skills in complementary areas. Also known as generalizing specialists.

task: A distinct element of scheduled work. *Also known as* activity.

task board: A display of project work that allows stakeholders to see the current status of tasks.

team charter: A document developed by the team that identifies team agreements and ways of working.

threat: A risk that will have a negative impact on a project.

threshold: A value that indicates action is required.

total float: The amount of time a task can slip and not affect a project constraint or the end date. Also known as float.

unknown risks: Risks you can't anticipate or plan for.

user story: A brief description of a desired outcome, documented from a stakeholder's perspective.

velocity: The rate at which a team accomplishes work in a time period.

value: Something of worth or importance.

waterfall: A predictive approach for creating deliverables that follows a linear pattern of completing one phase of work before starting the next one. *See also* predictive.

WBS dictionary: A document that provides a definition of work, activities, milestones, resources, costs, and other information for components of a WBS.

wideband Delphi: Working with a group of experts to develop individual estimates, discuss, and re-estimate until consensus is reached.

work breakdown structure (WBS): A tool used to decompose and organize project and product scope.

work package: The lowest level deliverable in a WBS.

Index

NOTE: Page references in *italics* indicate figures.

S